Modern Concepts of Peripheral Nerve Repair

Kirsten Haastert-Talini · Hans Assmus
Gregor Antoniadis

Editors

Modern Concepts of Peripheral Nerve Repair

Editors
Kirsten Haastert-Talini
Hannover Medical School
Institute of Neuroanatomy and Cell Biology
Hannover
Germany

Hans Assmus
Schriesheim
Germany

Gregor Antoniadis
Peripheral Nerve Surgery Unit
Department of Neurosurgery
University of Ulm
Günzburg
Bayern
Germany

ISBN 978-3-319-84875-4   ISBN 978-3-319-52319-4   (eBook)
DOI 10.1007/978-3-319-52319-4

Printed on acid-free paper

This Springer imprint is published by Springer Nature
The registered company is Springer International Publishing AG
The registered company address is: Gewerbestrasse 11, 6330 Cham, Switzerland

Preface

We, the editors of this book, join active membership of the German interdisciplinary study group named "NervClub" and we recognized that a concise book (comprisal) on modern concepts of peripheral nerve repair is currently not available for international readers. Therefore, we edited this book with a focus on very common and frequently occurring traumatic peripheral nerve injuries, their diagnostic with decision-making, and their reconstruction and long-term post-surgery patient care. This book should provide a compendium for graduated medical doctors interested in neurosurgery, hand surgery, or traumatology and final-year medical students with an upcoming interest in peripheral nerve surgery.

The topics have been carefully selected and the authors have treated them in a compact and illustrative way. The group of authors is comprised of internationally recognized experts in the field of peripheral nerve injury and repair from both the clinical and scientific points of view.

Our biggest thanks go to the authors for their enthusiasm to contribute to this project and to all those who helped us in editing this book.

We would like to thank further Ms. Lena Freund for her professional redrawing of our figures in Chaps. 1 and 10.

We thank Dr. Sylvana Freyberg, who, as the editor of Springer Heidelberg Medicine Books Continental Europe & UK, did accept our proposal and supported us through the project.

We also thank our project coordinator Ms. Rajeswari Balachandran and our project manager Ms. Govindan Meena, both from Springer Nature SPi Global, for their kind assistance.

<table>
<tr><td>Hannover, Germany</td><td>Kirsten Haastert-Talini</td></tr>
<tr><td>Schriesheim, Germany</td><td>Hans Assmus</td></tr>
<tr><td>Günzburg, Germany</td><td>Gregor Antoniadis</td></tr>
<tr><td>May 2017</td><td></td></tr>
</table>

Contents

The Peripheral Nerve: Neuroanatomical Principles Before and After Injury

1

Gregor Antoniadis

Contents

1.1   Anatomy of the Peripheral Nerve

Peripheral nerves arise from the spinal cord, and they contain axons from different types of neurons serving various effector organs or sensory endings.

The cell bodies of motor neurons that innervate skeletal muscle fibres are situated in the anterior horns of the grey matter of the spinal cord. Enlargements of the cord in the cervical and lumbar segments mark the major regions supplying the upper and lower limbs (Fig. 1.1).

The first sensory neurons are situated in the dorsal root ganglia, which are located in the intervertebral foramina, just proximal to the fusion of the anterior and posterior roots.

The peripheral nerve is composed of motor, sensory and sympathetic nerve fibres. A nerve fibre is the conducting unit of the nerve and contains the following elements: *a central core, the axon and Schwann cells.* Some nerve fibres are surrounded by a myelin component (*myelinated nerve fibres*), and others are free of such myelin sheath (*unmyelinated nerve fibres*) [2, 9, 17, 20].

G. Antoniadis, MD, PhD
Peripheral Nerve Surgery Unit, Department of Neurosurgery, University of Ulm,
Ludwig-Heilmeyer-Str. 2, 89312 Günzburg, Germany
e-mail: gregor.antoniadis@uni-ulm.de, pnc.antoniadis@gmail.com

© Springer International Publishing AG 2017
K. Haastert-Talini et al. (eds.), *Modern Concepts of Peripheral Nerve Repair*,
DOI 10.1007/978-3-319-52319-4_1

1

The axons contain organelles including mitochondria, neurofilaments, endoplasmic reticulum, microtubules and dense particles. Axons originate from their corresponding neuronal cell bodies which are located in the spinal cord, dorsal root ganglia or autonomic ganglia, respectively.

The Schwann cells are the glial cells of the peripheral nervous system and located along the longitudinal extent of the axon.

In healthy peripheral nerves, nerve fibres of different diameter exist, large and small fibres. Only *large fibres* (>1.5 µm in diameter) are surrounded by segmental lipoprotein coating or covering of myelin. In this case the membranes of neighbouring Schwann cells wrap concentrically around a segment of the axon. The small area between the neighbouring Schwann cells is known as the "node of Ranvier". The node of Ranvier permits ionic exchanges between the axoplasm of a nerve fibre and the intercellular space and permits saltatory conduction of a nerve action potential impulse, which jumps from one node to the next and is the basis for fast signal conduction. There is a basal lamina around each Schwann cell and its contents (Fig. 1.1 and 1.2).

The small and less myelinated fibres (<1.5 µm in diameter) are often grouped and enveloped by the membrane of a Schwann cell, which does not wrap a lipoprotein sheath around them (Remak bundles). These fibres do not have the structural capacity for saltatory conduction, and nerve impulses transmit slowly along the axon.

A Schwann cell not only provides myelins and a basal membrane as guidance for axons but, as a source of trophic and growth factors, it supports also the maintenance of its neighbouring axon.

The connective tissue which forms the supporting framework for the nerve fibres is the *interfascicular endoneurium*. A thin sheath of specialized perineurial cells, called *perineurium*, covers a bundle of nerve fibres (*fascicle*).

The nerve fascicles vary in number as well as in size, depending on a given nerve as well as the level of the nerve examination.

The endoneurium is a matrix of small-diameter collagen fibrils which are predominantly longitudinally oriented. Microvessels with tight junctions are found at this structure, and the tissue adjacent to these capillaries probably serves as a blood-nerve barrier additional to the endoneurial tissue itself [2, 7].

The perineurium consists of oblique, circular and longitudinal collagen fibrils dispersed amongst perineurial cells [23]. The outer lamellae of the perineurium have a high density of endocytotic vesicles which may play a role in molecular transport, e.g. of glucose. The inner lamellae have tight junctions between contiguous perineurial cells, which may block the intercellular transport of macromolecules and crucially contribute to a blood-nerve barrier [10]. The interruption of the perineurium can affect the function of the axons, which it encloses. The perineurium is the major source of tensile strength for nerve and is transversed by vessels which carry a perineurial sleeve of connective tissue.

The epineurium represents the connective tissue that covers the entire nerve trunk. The *epineurium* can extend internally to separate the fascicles (*interfascicular epineurium*). The layer between the epineurium and the surrounding tissue is called paraneurium.

The axoplasm contains proteins and cytoskeletal elements including microtubules and neurofilaments. The axoplasm is continuously built and sustained by axonal transport mechanisms.

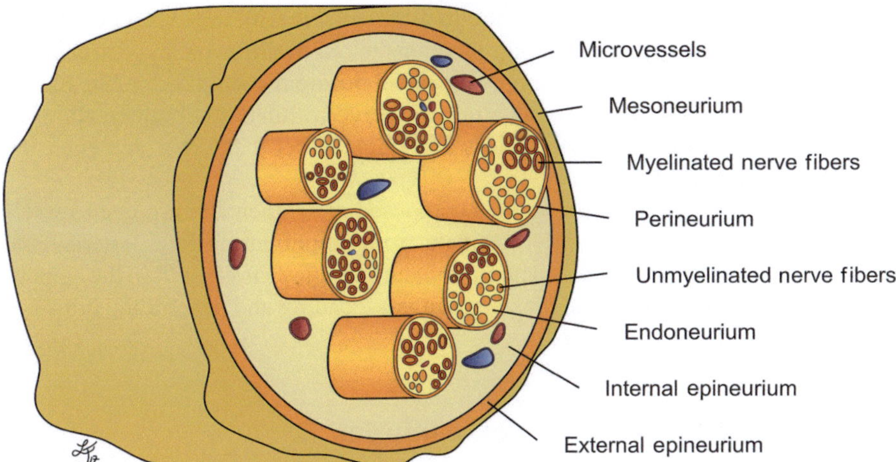

Fig. 1.1 Schematic diagram of a normal nerve (Illustration by Lena Julie Freund, Aachen, Germany)

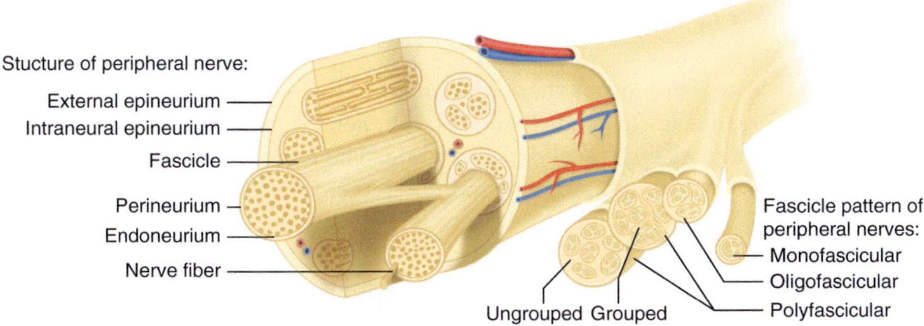

Fig. 1.2 Morphology of the peripheral nerve (From Kretschmer et al. [9])

The relationship between the fascicles within the peripheral nerve is constantly changing along a longitudinal course. Sunderland noted that the maximum length of nerve with a constant pattern was 15 mm [22].

Three types of nerves concerning their fascicular pattern can be distinguished [12, 13]:

1. Nerves with a monofascicular pattern.
2. Nerves with an oligofascicular pattern (2–10 fascicles).
3. Nerves with a polyfascicular pattern. For this nerve type, there are two subtypes that can be distinguished: the polyfascicular nerve with diffuse arrangement of fascicles and the polyfascicular nerve with group arrangement of fascicles.

Peripheral nerves receive the *blood supply* from small vessels leading to the epineurium (intrinsic), perineurium and endoneurium. The normal nerve is critically dependent upon the intrinsic blood supply and the perineurial and endoneurial vessels. The *intrinsic vessels* are similar to other vessels with the exception of having endothelial cells that contain tight junctions to aid in diffusion and extrusion of compounds. The intrinsic blood supply is crucial during regeneration, as the blood-nerve barrier breaks down uniformly along the nerve within days of injury, allowing large molecules, such as growth factors and immune cells, to cross and enter the endoneurial space [16].

The *extrinsic blood supply* system is composed of segmentally arranged vessels which vary in size and generally originate from neighbouring large arteries and veins. As these nutrient vessels reach the epineurium, they ramify within the epineurium and supply the intraneural plexus through ascending and descending branches [11, 12].

1.2 Classification of the Nerve Injuries

More than 70 years ago, nerve lesions were characterized as compression, contusion, laceration or division lesions. *Seddon* introduced in 1943 a classification system based on nerve fibre and nerve trunk pathology in three categories: neurapraxia, axonotmesis and neurotmesis [21].

The new classification according to *Sunderland* is based on histological features of the nerve trunk in 5° [18, 22] (Figs. 1.3 and 1.4):

Grade I By this type of lesion, there is an interruption of conduction at the site of injury. Therefore, this lesion grade corresponds to neurapraxia of Seddon classification. It is characterized by a focal demyelination. The rearrangement of the myelin sheath takes 3–4 weeks. After this period the nerve can almost regain its normal function.

Grade II This lesion corresponds to Seddon axonotmesis. The axon is severed, but the endoneurial sheath of nerve fibre and the basal lamina are preserved. The axons undergo Wallerian degeneration. The regeneration process lasts some months, depending on the distance between lesion and target muscle. The regeneration process may result nearly in a restitutio ad integrum.

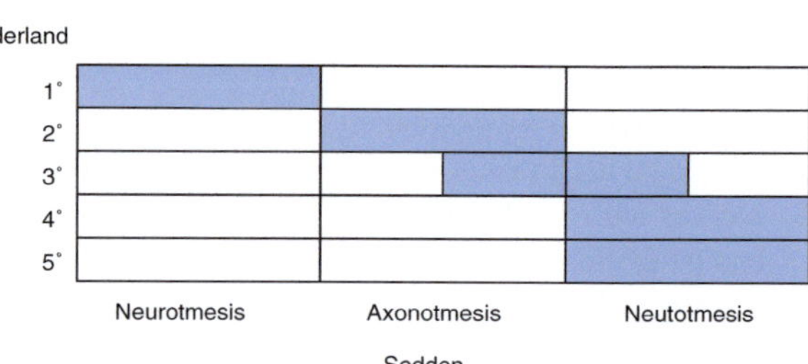

Fig. 1.3 Correlation of classifications according to Seddon and to Sunderland

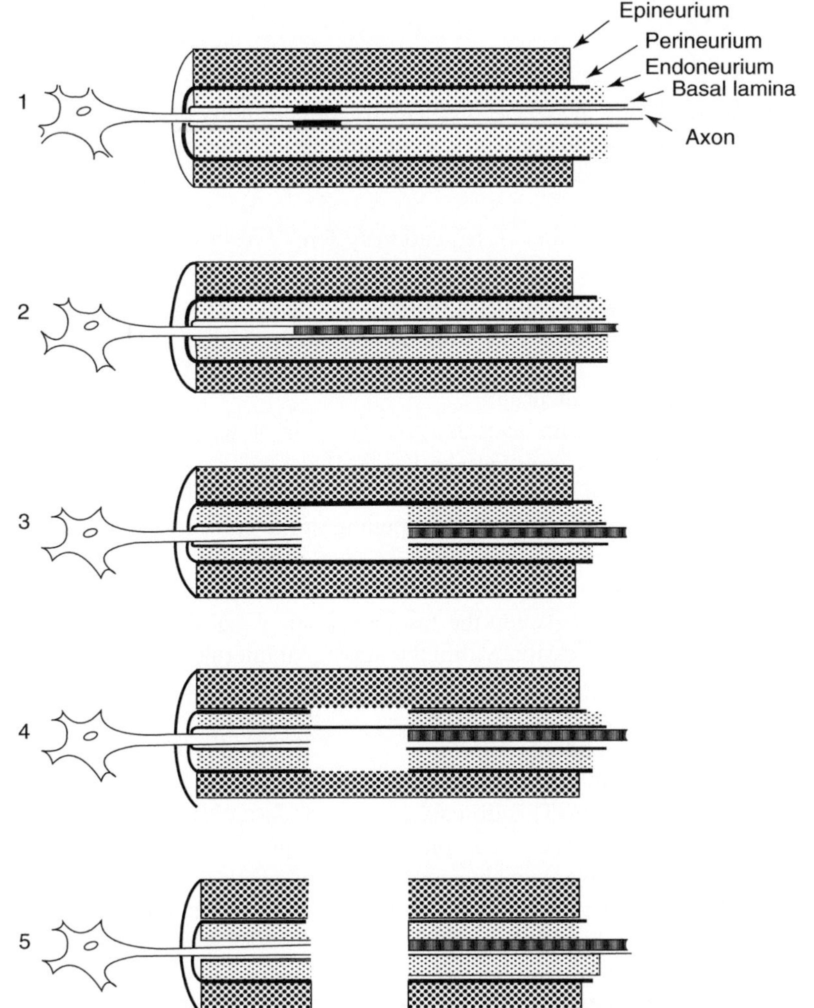

Fig. 1.4 Classification according to Sunderland (With permission from Terzis and Smith [25])

Grade III The essential features of these injuries are destructions of endoneurial structures of the nerve fibres. A disintegration of axons and Wallerian degeneration and loss of the endoneurial tube continuity occur. The perineurium is kept intact. This situation leads to a certain degree of misdirection of regenerating axons, followed by extensive unrecoverable functional deficits.

Grade IV In grade IV injuries, the fasciculi and the perineurium are ruptured, while the epineurium is still preserved. Compressing forces are even able to block the outgrowth of regeneration axon sprouts, resulting in a neuroma in continuity.

The lesion grades IV and V present a very poor prognosis concerning spontaneous recovery.

Grade V In this lesion there is a complete loss of continuity of the nerve trunk and no chance for a spontaneous recovery. Regenerative attempts produce neuroma formation on the separated nerve stumps.

Millesi established a further classification system of reactive nerve fibrosis types. The connective tissue reaction upon lesion surrounding the nerve fibre, fascicles and fascicle groups is designated in these cases as fibrosis [13–15].
Millesi classified the fibrosis in three types:

Type A: Fibrosis of the epineurium.
The thickened epineurium leads to a strangulation of nerve structures. This type of fibrosis can occur in all cases of grades I, II or III lesions, according to Sunderland's classification. Chances of spontaneous recovery remain decreased as usually a long-lasting nerve compression occurred. An opening of the epineurium (epineurotomy) is the treatment of choice.
Type B: Fibrosis of the epifascicular and interfascicular epineurium.
The connective tissue between the fascicles is involved. Each fascicle group is affected. There is a compression within the nerve. An internal neurolysis to decompress all fascicle groups is indicated. This procedure must be done meticulously and using the microscope.
Type C: Fibrosis of the endoneurium.
This type of fibrosis takes place in grades III and IV lesions according to Sunderland's classification. This fibrosis involves all fascicles and presents with the danger for the development of a neuroma in continuity. Type C fibrosis has a very poor prognosis, and there are no chances for a spontaneous recovery. A reconstruction of the nerve after the resection of damaged parts must be done.

1.3 Neuroanatomical Situation After Injury: Nerve Degeneration

Axons, Schwann cells, macrophages, fibroblasts and other cell types demonstrate significant changes in response to nerve injury.
An injury of the neuronal soma, in very proximal lesions, is a very severe nerve injury without potential for recovery. It occurs in injuries with direct mechanical or vascular insult to the neuronal soma [8].
In peripheral axonal injuries, neuronal cell death does not occur, in contrast to an avulsion of the nerve roots, which results in neuronal cell death and loss of the soma. Therefore, peripheral nerves have a regenerative potential after injury.
Waller described in 1850 an antegrade nerve degeneration (*Wallerian degeneration*), which is characterized by a loss of cellular integrity and trafficking of intracellular components along the distal nerve end. We know today that a degeneration of the neuromuscular synapses can precede the process for several hours and that this is independent from the Wallerian degeneration [5].

The process of Wallerian degeneration is completed after several weeks, and it includes the gradual dissolution of axoplasm and myelin distal to an injury and their gradual phagocytosis or their debris (Fig. 1.5).

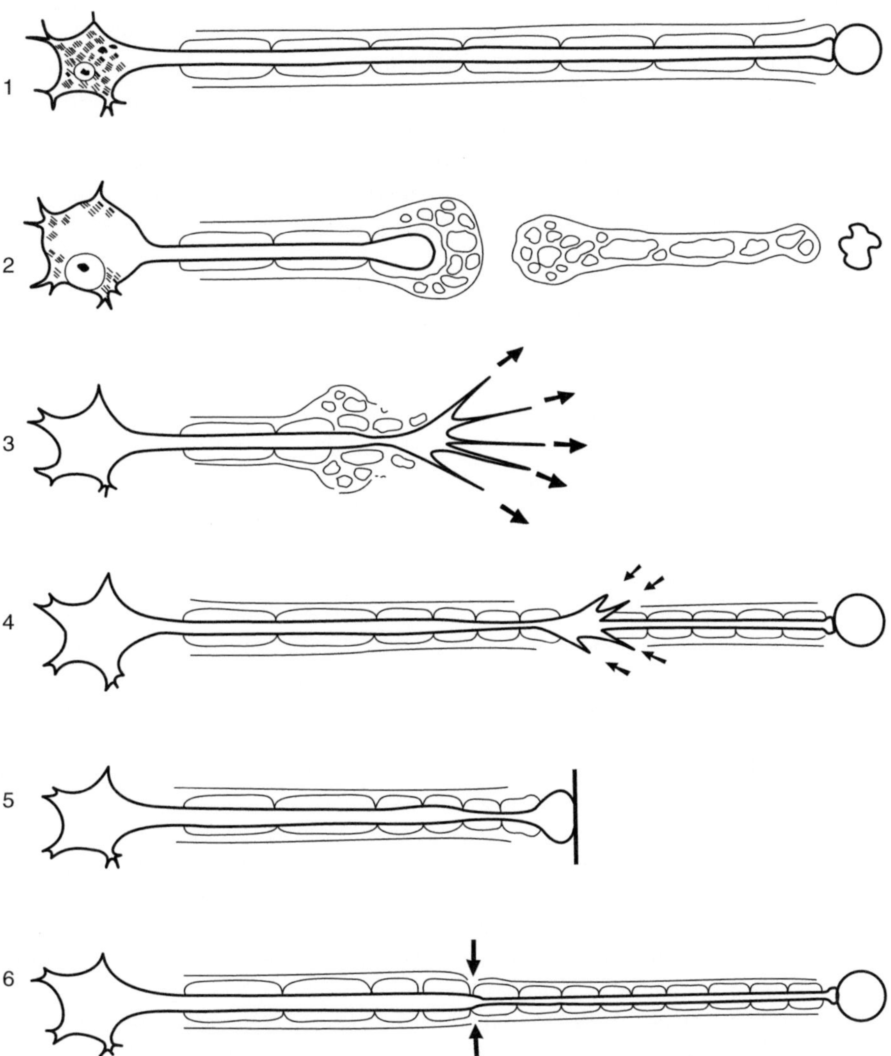

Fig. 1.5 Degeneration and regeneration after nerve injury (With permission from Terzis and Smith [25]) (*1*) Normal, healthy neuron with myelinated axon. (*2*) After transection injury, the proximal axon has undergone retrograde reaction with somal chromatolysis, nuclear migration and nuclear enlargement. The distal axon has undergone Wallerian degeneration. (*3*) The proximal axon has begun to sprout filopodia from the growth cone to begin axonal regeneration. (*4*) Upon successfully re-establishing connectivity of one axonal sprout, redundant sprouts undergo the dying-back process. (*5*) A terminal bulb is created by damming up axonal contents if an insurmountable obstruction is encountered. (*6*) If scar produces an annular constriction around a regenerating axon (*arrows*), the resulting axonal calibre will never return to normal [26]

Observations of Wallerian degeneration have revealed that the initial degradation of axonal components rapidly leads to recruitment and activation of non-neuronal cells that crucially contribute to the regeneration process. This process includes the dedifferentiation, proliferation and migration of Schwann cells, with the activation of macrophages within the endoneurium and the recruitment of complementary immune cells from the periphery. These cells prepare the distal nerve for regeneration by clearing myelin debris and other inhibitors to axonal regeneration during the neural wound-healing response.

The recruitment of Schwann cells and macrophages is linked to the secretion of several other specific pro-inflammatory cytokines and chemokines. The secretion of these pro-inflammatory agents similarly stimulates the recruitment of macrophages and immune-competent cells from the periphery [19].

1.4   Nerve Regeneration

As a consequence of any severe injury to a peripheral nerve, there is a predictable sequence of distal and proximal axonal degeneration. Whenever the injury does not lead to neuronal death, a sequence of regeneration proceeds, which may be abortive or may result in effective functional restoration.

Proximal to the axonal disruption, the axon undergoes limited degeneration up to the last preserved internode (node of Ranvier). This axonal degeneration is similar to that observed in the distal stump. The neuron exhibits central chromatolysis, and this represents the metabolic preparation for a shift from maintenance of nerve conduction to a regenerative mode reprogrammed to generate structural proteins.

Ramon-y-Cajal started the modern era in nerve regeneration research by proving that nerve regeneration occurs by axonal outgrowth from the proximal stump and not by autoregeneration of the degenerated distal nerve [3].

During 24 h after transection injury, the proximal axon bulges into a growth cone. By the end of the first 24 h, a few sprouts have reached the areas of injury, and the penetration of the developing scar at the site of injury proceeds from the second to third day. The axonal sprouts originating from the proximal nerve stump are accompanied by Schwann cells derived from the reciprocation of the terminal satellite cells. The growth cone is rich of endoplasmic reticulum, microtubules, microfilaments, large mitochondria, lysosomes and other vacuolar and vesicular structures of unknown significance.

During Wallerian degeneration of the distal nerve end, Schwann cells assume the dual role of phagocytosis of myelin and axonal debris, and they proliferate within the basal lamina of remaining endoneurial connective tissue sheaths. As they proliferate, Schwann cells become densely packed in longitudinal rows histologically recognized as the *bands of Bungner* [6].

After a peripheral nerve injury, three types of intrinsic and extrinsic processes affect the neuron, on molecular level:

1. Positive signals, derived from retrograde transport of kinases, such as mitogen-activated protein kinases (MAPKs), are transported from the injury site to the cell body [1].

2. Axonal injury leads to the disruption of action potential conduction and a large influx of calcium and a depolarizing wave. This initial calcium influx leads to protein kinase C (PKC) activation within the cell body and the nuclear export of a regeneration-associated gene repressor [4].
3. Interruption of retrograde transport of trophic factors and negative regulators of axonal growth from the end organ leads to the upregulation of regeneration-associated genes [1].

Current research on the intrinsic regeneration capacity of neurons focuses on the interplay between cytoskeletal assembly and blocking of the inhibitory effects of myelin.

A regenerating axon grows through a scar with an average rate of about 0.25 mm per day, and once the axonal sprouts reach the distal endoneurial tube, axonal growth continues at an average 1.0–8.5 mm per day, depending upon multiple factors. The speed of nerve regeneration is inversely proportional to the distance of the nerve injury from the cell body as observable by the progressing Tinel's sign. In each instance the nerve will be regenerating under ideal conditions. When the underlying bed is well vascularized, nerve regeneration proceeds through non-vascularized nerve grafts at 2–3 mm per day, and axonal elongation is even faster through vascularized nerve grafts, approaching 3–4 mm per day [24, 25].

It is recognized that the quality and speed of regeneration of a nerve are improved when there is a minimum amount of scar tissue filling the gap. Also, a better prognosis for return of function exists when regenerating axons enter their native endoneurial tubes and become guided back towards the appropriate target organ. Beside pure mechanical factors, the amount of time, which is allowed to elapse between injury and repair, is very important for the prognosis of functional recovery. Other factors such as the age of the patient, the type of nerve, the level of nerve injury, the cause of the injury and the associated injuries all influence the functional outcome after nerve reconstruction, but cannot yet be manipulated.

References

1. Abe N, Cavalli V. Nerve injury signaling. Curr Opin Neurobiol. 2008;18:276–83.
2. Birch R. Surgical disorders of the peripheral nerves. 2nd ed. London: Springer; 2011.
3. Cayal RY. Degeneration and regeneration of the nervous system, vol. 1. London: Oxford University Press; 1978.
4. Cho Y, Sloutsky R, Naegle KM, Cavalli V. Injury-induced HDAAC5 nuclear export is essential for axon regeneration. Cell. 2013;155:894–908.
5. Gillingwater TH, Ribschester RR. The relationship of neuromuscular synapse elimination to synaptic degeneration and pathology: insights from WldS and other mutant mice. J Neurocytol. 2003;32:863–81.
6. Jessen KR, Mirsky R. The repair Schwann cell and its function in regenerating nerves. J Physiol. 2016;594:3521–31.
7. Kim DH, Midha R, Murovic JA, Spinner RJ. Kline and Hudson's nerve injuries. Philadelphia: Saunders Elsevier; 2014.
8. Koliatsos VE, Price WL, Pardo CA, Price DL. Ventral root avulsion: an experimental model of death of adult motor neurons. J Comp Neurol. 1994;342:35–44.
9. Kretschmner T, Antoniadis G, Assmus H. Nervenchirurgie. Heidelberg: Springer Verlag; 2014.

10. Lundborg G, Rydevick B. Effects of stretching the tibial nerve of the rabbit. A preliminary study of the intraneural circulation and the barrier function of the perineurium. J Bone Joint Surg Br. 1973;55B:390–401.
11. Lundborg GA. A 25-year perspective of peripheral nerve surgery: evolving neuroscientific concepts and clinical significance. J Hand Surg Am. 2000;25:391–414.
12. Mackinnon S. Nerve surgery. New York: Thieme Medical; 2016.
13. Millesi H. Chirurgie der peripheren Nerven. München: Urban & Schwarzenberg Verlag; 1992.
14. Millesi H. Reappraisal of nerve repair. Surg Clin North Am. 1981;61:321–40.
15. Millesi H, Rath TH, Reihsner R, Zoch G. Microsurgical neurolysis: its anatomical and physiological basis and its classification. Microsurgery. 1993;14:430–9.
16. Olson Y. Studies on vascular permeability in peripheral nerves. I. Distribution of circulating fluorescent serum albumin in normal, crushed and sectioned rat sciatic nerve. Acta Neuropathol. 1966;7:1–15.
17. Penkert G, Fansa H. Peripheral nerve lesions. Heidelberg/New York: Springer Verlag; 2004.
18. Penkert G, Boehm J, Schelle T. Focal peripheral neuropathy. Heidelberg: Springer Verlag; 2015.
19. Perrin FE, Lacroix S, Avilés-Trigueros M, David S. Involvement of monocyte chemoattractant protein-1, macrophage inflammatory protein-1alpha and interleukin-1beta in Wallerian degeneration. Brain. 2005;128:854–66.
20. Rea P. Essential clinically applied anatomy of the peripheral nervous system in the limbs. Amsterdam: Elsevier; 2015.
21. Seddon HJ. Three types of nerve injury. Brain. 1943;66:237–88.
22. Sunderland S. Nerve injuries and their repair. A critical appraisal. Edinburgh/London/Melbourne/New York: Churchill Livingstone; 1991.
23. Spencer PS, Weinberg HJ, Paine CS. The perineurial window- a new model of focal demyelination and remyelination. Brain Res. 1975;965:923–9.
24. Terzis JK. Microreconstruction of nerve injuries. Philadelphia: Saunders Company; 1987.
25. Terzis JK, Smith KL. The peripheral nerve – structure function reconstruction. New York: Raven Press; 1990.
26. Weiss P. The technology of nerve regeneration: a review. Sutureless tubulation and related methods of nerve repair. J Neurosurg. 1944;1:400–50.

State-of-the-Art Diagnosis of Peripheral Nerve Trauma: Clinical Examination, Electrodiagnostic, and Imaging

2

Christian Bischoff, Jennifer Kollmer,
and Wilhelm Schulte-Mattler

Contents

2.1　Introduction

Peripheral nerve trauma is no exception from the rule that appropriate treatment requires a clear diagnosis. Specific clinical diagnostic tests, such as Hoffmann and Tinel's sign, and technologies, such as electrodiagnostic and imaging procedures,

C. Bischoff (✉)
Neurologische Gemeinschaftspraxis, Burgstraße 7, D-80331 Munich, Germany
e-mail: bischoff@profbischoff.de

J. Kollmer (✉)
Department of Neuroradiology, University of Heidelberg,
Im Neuenheimer Feld 400, D-69120 Heidelberg, Germany
e-mail: jennifer.kollmer@med.uni-heidelberg.de

W. Schulte-Mattler (✉)
Department of Neurology, University Hospital Regensburg/Bezirksklinikum Regensburg,
Universitätsstraße 84, D-93053 Regensburg, Germany
e-mail: wilhelm.schulte-mattler@klinik.uni-regensburg.de

© Springer International Publishing AG 2017

11

K. Haastert-Talini et al. (eds.), *Modern Concepts of Peripheral Nerve Repair*,
DOI 10.1007/978-3-319-52319-4_2

were developed to improve diagnostic accuracy [19]. The development has not yet come to an end. In the recent years, major accomplishments were achieved in neuroimaging methods of peripheral nerves [22, 29]. The more classical electrodiagnostic methods benefitted from a better understanding of the timing of the pathological findings after an injury [11, 17].

2.2 History Taking and Physical Examination

In a traumatized patient, careful history taking and physical examination are mandatory, as they provide the key information to answer the following questions:

- Is patient's pain nociceptive, or neuropathic, or both?
- Are there neurological deficits that can be attributed to one or to multiple lesions of peripheral nerves?
- Where are the lesions?
- How severe and of what type (neurapraxia, axonotmesis, neurotmesis, see Chap. 1) are the lesions?
- Are there neurological deficits that cannot necessarily be explained by a peripheral nerve lesion but, for instance, by a spinal or by a cerebral lesion?
- Are there pre-existing pathological conditions, such as peripheral neuropathy, that contribute to the patient's actual signs and symptoms?

Only in rare cases, all of these questions can sufficiently be answered on history and clinical examination alone. In the other cases, history and clinical examination constitute the basis for rational decisions about necessity and timing of additional diagnostic testing, especially electrodiagnostic and imaging studies, which are of established high value.

The answers to the questions listed above provide the information to sensibly make therapeutic decisions (see also Chap. 3).

During follow-up after a nerve trauma, the key question is whether reinnervation takes place in time or not. This can be assessed clinically, as it is nicely illustrated by the original descriptions of what nowadays is known as Hoffmann-Tinel's sign. Paul Hoffmann and Jules Tinel independently from each other described the occurrence of tingling by pressure applied to an injured nerve. Hoffmann shortly later reported that the paresthesia could also be elicited by percussion of the nerve.

They used their observation to monitor nerve regeneration after its surgical repair [15, 36]. Nonetheless, in many clinical situations, electrodiagnostic and imaging studies may be necessary to provide additional relevant information.

Finally, the functional outcome is judged mainly clinically. Again, electrodiagnostic and imaging studies may provide necessary informations, which not infrequently are relevant for forensic purposes.

2.3 Electrodiagnostic Procedures

The most relevant electrodiagnostic methods to assess peripheral nerve lesions are needle electromyography (EMG) and nerve conduction studies (NCS) [6, 8, 19]. For an EMG study, a needle electrode is inserted into the target muscle, and its electrical activity at rest and at various degrees of voluntary contraction is recorded from multiple positions within that muscle. The idea behind EMG is that axonal damage causes functional and structural changes of the motor units of the innervated muscle, which consequently result in changes of the electrical properties of the affected motor units.

For an NCS a peripheral nerve is stimulated electrically, and the resulting action potentials are recorded. The idea behind NCS is that a loss of functional axons causes a loss of amplitude of the recorded action potentials. Recordings can be made from the nerve itself (sensory or mixed NCS) or from a muscle innervated by that nerve (motor NCS). The diagnostic yield of a motor NCS is much higher if the stimulation is done not only at one but at two or more sites along that nerve and if the action potentials of different stimulation sites are compared with each other. Tables 2.1 and 2.2 list pathological findings and their diagnostic meaning.

The tables indicate that the electrodiagnostic findings and their time course may provide valuable information about both the site and the type of a suspected nerve lesion. Table 2.3 is intended to help the reader plan a sensible timing for the diagnostic tests and to "decode" their results.

Some time intervals after the trauma are noteworthy [8].

Table 2.1 EMG findings in a weak muscle and their diagnostic meaning

EMG finding	Occurrence/cause	Timing
Normal	Central nervous system lesion (e.g., intracranial hemorrhage)	Always
	Not a severe nerve lesion	Until the occurrence of pathological spontaneous activity
Pathologic spontaneous activity	Axonotmesis, neurotmesis, myopathy	Begins 10–14 days after a lesion, ends after full recovery, may persist for decades if recovery is incomplete
Polyphasic motor unit action potentials (MUAPs)	Partial axonotmesis, myopathy	Begins 6 weeks after an incomplete lesion, ends after recovery, some may persist
Large MUAPs (may be polyphasic)	Partial axonotmesis	Begins 6–12 months after an incomplete lesion, persists after recovery
Increased (>20/s) discharge rate of single motor units	Neurapraxia, partial axonotmesis	Begins immediately after the lesion, accompanies weakness

Table 2.2 NCS findings in a nerve supplying a weak muscle and their diagnostic meaning

NCS finding	Occurrence/cause	Timing
Normal compound muscle action potentials (CMAPs)	Central nervous system lesion (e.g., intracranial hemorrhage),not a severe nerve lesion, myopathy	Always
Normal (CMAPs)	Stimulation distal to the lesion only	Ends 4–7 days after the lesion (due to Wallerian degeneration of axons)
Normal CMAPs upon nerve stimulation distal to the lesion, low CMAPs upon proximal stimulation[a]	Neurapraxia (also called "conduction block")	Begins with the lesion, ends with recovery
Normal CMAPs upon nerve stimulation distal to the lesion, low CMAPs upon proximal stimulation[a]	Axonotmesis	Ends 4–7 days after the lesion (due to Wallerian degeneration of axons)
Normal CMAPs upon nerve stimulation distal to the lesion, low CMAPs upon proximal stimulation[a]	Innervation anomaly	Always
Low CMAPs at all stimulation sites, low sensory nerve action potentials (SNAPs)	Axonotmesis	Begins 4–7 days after the lesion (due to Wallerian degeneration of axons), ends with full recovery
Mildly reduced nerve conduction velocity (NCV) (leg, 30–40 m/s; arm, 40–50 m/s)	Axonotmesis	Parallels low CMAPs
Mildly reduced nerve conduction velocity (NCV) (leg, 30–40 m/s; arm, 40–50 m/s)	Pre-existing polyneuropathy	No relationship to the nerve trauma
Severely reduced NCV (leg, <30 m/s; arm, <40 m/s)	Demyelinating neuropathy, not caused by nerve trauma	No relationship to the nerve trauma

[a]Note that this finding is often labeled "conduction block," although conduction block is only one of its potential causes

Immediately after a severe lesion an EMG may be valuable: If motor unit action potentials (MUAPs) are recorded, the lesion is incomplete, and thus neurotmesis is ruled out. If pathologic spontaneous activity (PSA) is recorded *within the first 10 days* or if abnormally polyphasic or enlarged MUAPs are found *within the first 4 weeks*, a pre-existing neuropathy (or, rarely, a myopathy) is documented. It should also be noted that PSA may persist for years. Thus, the occurrence of PSA not necessarily indicates a recent lesion. The recency of a lesion can be inferred from PSA only if the PSA was not found in an early recording but does appear later on.

If increased discharge rates of motor units are found *at any time*, a central nervous system lesion is ruled out.

An NCS may make particular sense *within the first 4 days*, namely, before Wallerian degeneration (see Chap. 1) becomes apparent [11]. Only during this time, the distal part of the lesioned nerve can be stimulated electrically. This results in a

Table 2.3 Electrodiagnostic findings after a nerve trauma over time

Time after trauma		Type of lesion		
		Neurapraxia	Partial axonotmesis	Total axonotmesis, neurotmesis
Immediately	EMG	No PSA, DR ↑ MUAPs n	No PSA, DR ↑, MUAPs n	No PSA, no MUAPs
	NCS	ΔCMAP	ΔCMAP	ΔCMAP
4–7 days	EMG		No PSA, DR ↑, MUAPs n	No PSA, no MUAPs
	NCS		CMAPs ↓	No CMAPs
10–20 days	EMG		PSA, DR ↑, MUAPs n	PSA, no MUAPs
	NCS		CMAPs ↓	No CMAPs
>6 weeks	EMG	n	PSA, DR ↑, polyphasic MUAPs	PSA, small polyphasic ("nascent") MUAPs
	NCS	n	CMAPs ↓	No CMAPs
Years	EMG	n	MUAPs ↑	(PSA), MUAPs ↑
	NCS	n	CMAPs (↓)	CMAPs ↓

DR discharge rate (of motor units!), *MUAP* motor unit action potential, *PSA* pathologic spontaneous activity, *CMAP* compound muscle action potential, *ΔCMAP* normal CMAPs upon nerve stimulation distal to the lesion, low CMAPs upon proximal stimulation (see Table 2.2), *n* normal, ↑ pathologically increased, ↓ pathologically decreased

normal compound muscle action potential (CMAP) following stimulation distal to the lesion and a low CMAP following stimulation proximal to the lesion, a finding that reliably localizes the nerve lesion. After the completion of the Wallerian degeneration, namely, *after 11 days* [11], all CMAPs are low or have disappeared, irrespective to where the lesion is located. If a low CMAP upon distal stimulation is found *within the first 4 days*, a pre-existing lesion is documented. Conversely, a normal CMAP during that time documents the integrity of that nerve before the trauma. This finding, as well as the absence of PSA early on EMG, can be particularly helpful in the evaluation of potentially iatrogenic nerve lesions.

A major shortcoming of the established electrodiagnostic methods is that they do not help to make the important distinction between neurotmesis and total axonotmesis; the latter denominates a condition of a nerve characterized by all of its axons suffering from axonotmesis.

Figure 2.1 illustrates a common clinical situation and how a good knowledge about the benefit of electrodiagnostic procedures and the meaning of the respective findings are important for appropriate treatment decisions.

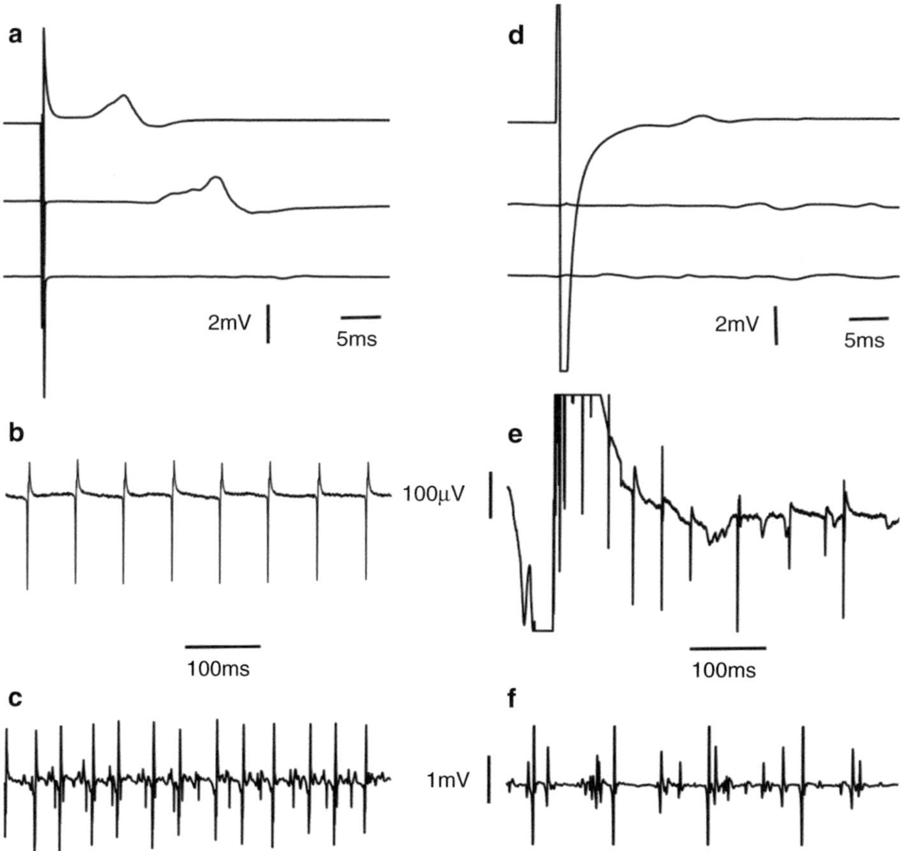

Fig. 2.1 A 74-year-old man experienced plegia of his left foot extensors immediately after surgery on his lumbar spine. As a complication of the surgery was suspected, the patient underwent a second operation 1 day after the first one, which did not resolve the problem. First electrodiagnostic examination was done 2 days after the first surgery: (**a**) motor nerve conduction study (NCS) recordings from his left extensor digitorum brevis muscle, stimulation of the peroneal nerve at the dorsum of the foot (upper trace), and below and above the fibular head (lower traces). Compound muscle action potentials (CMAPs) upon distal stimulation are low, indicating a pre-existing lesion, and CMAPs upon proximal stimulation are absent, which shows that there is an additional lesion that can be localized at the fibular head (Table 2.3, "immediately"). (**b**) The electromyogram (EMG) of the anterior tibial muscle shows pathologic spontaneous activity (PSA), which also demonstrates a pre-existing lesion. (**c**) Increased (>20/s) discharge rates of motor units show that at least 80% of the motor units of the muscle are not functional [31]. These results point to the site of the actual lesion and show the pre-existing one. The type of the lesion cannot be inferred. A subsequent electrodiagnostic examination was done 20 days after surgery: (**d**) NCS as in (**a**) all CMAPs are absent, showing that the type of the lesion is axonotmesis (or neurotmesis) (Table 2.3, "10–20 days"). (**e**) The electromyogram (EMG) of the anterior tibial muscle shows pathologic spontaneous activity (PSA), showing that the type of the lesion is axonotmesis that took place at least 10–14 days before this recording was made. (**f**) Discharge rates of motor units are normal, showing a functional recovery of many motor units of this muscle since the recording (**c**) was made. These results do not permit to localize the lesion but show that the lesion type is partial axonotmesis, more pronounced in the extensor digitorum brevis than in the anterior tibial muscle. It should be noted that the second operation could have been avoided if the first electrodiagnostic examination had taken place immediately

2.4   Imaging

2.4.1   High-Resolution Ultrasound

Ultrasound imaging of peripheral nerves is done since a quarter of a century [13]. Initially, this was done virtually exclusively by radiologists and orthopedic surgeons. Neurological studies on this subject were published from the beginning of this millennium [2]. Since then technology had made extreme progress, especially the spatial resolution of ultrasound was dramatically improved. However, the penetration depth of ultrasound is still limited. This is the main shortcoming of ultrasound, especially if compared with magnetic resonance imaging (MRI) [29]. As a consequence of the methodological improvements, the number of publications on "ultrasound" and "peripheral nerve" increased from less than one per year before 2000 to 60 PubMed entries in 2015.

To date, the clinical significance of ultrasound imaging of peripheral nerves in general is not without controversial discussion [5], while its role in the diagnosis of traumatic nerve lesions is yet better defined [7, 20, 27, 37]. This is because the major diagnostic issue in traumatic lesions is to determine both the type and the morphology of the lesion and not so much to localize the lesion. As a major point, the important distinction between neurotmesis and total axonotmesis, which cannot be made with electrodiagnostic methods, can readily be made with ultrasound. When the diagnostic value of ultrasound was studied prospectively in 65 patients with nerve trauma, the use of ultrasound strongly modified the diagnosis and the therapy in 58 % of cases. It specifically contributed to the following:

- Distinction between neurotmesis and axonotmesis
- Identification of etiology
- Demonstration of multiple sites of nerve damage

The contribution of ultrasound was clearly the highest in cases with neurophysiological evidence of complete axonal damage [27]. Figure 2.2 illustrates a typical clinical situation in which ultrasound imaging clearly demonstrates a peripheral nerve's neurotmesis.

Ultrasound can be used to study the development of neuromas, both before and after nerve surgery. Unfortunately, the information that can be drawn from such imaging is of limited value so far, as there is no relation between enlargement of neuroma and nerve function unless the size of the neuroma exceeds a cutoff beyond which prognosis is negative [9].

Before nerve surgery, ultrasound can be used to detect the location of proximal and distal nerve stumps. They can be marked on the skin preoperatively to help the surgeon better tailor the procedure to the damaged nerve's needs and save time that otherwise would be needed for the search for the stumps [20].

During nerve surgery, ultrasound imaging can time efficiently and reliably be used to assess the severity of the underlying nerve injury and the type (intraneural/perineural) and grade of nerve fibrosis [21].

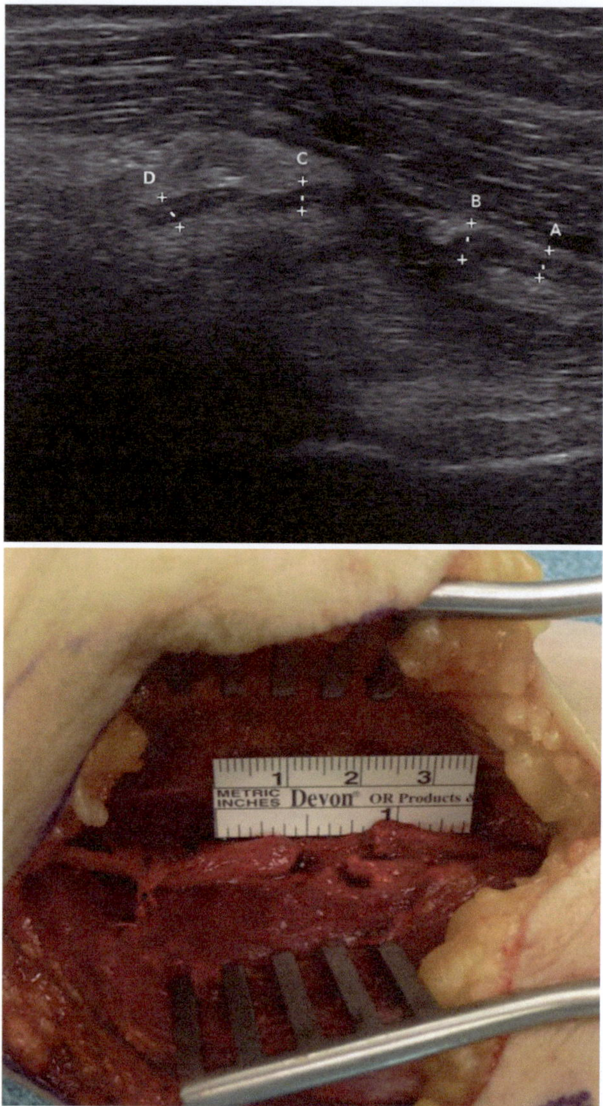

Fig. 2.2 A 64-year-old woman got a lipoma removed from her cubital fossa. Immediately after surgery she experienced plegia of her finger extensors. Four weeks after, there still was plegia of all muscles innervated by her radial nerve distal to the extensor carpi radialis brevis muscle. Upon EMG examination of the plegic muscles, there was abundant spontaneous activity but no MUAPs. High-resolution ultrasound imaging (*upper*, provided by Peter Pöschl, Regensburg) clearly shows a transected nerve, with (*A, B*) and (*C, D*) marking the nerve stumps. This finding is confirmed by visual inspection during subsequent surgery (*lower*)

If carried out monthly after nerve surgery, ultrasound examinations could earlier pick up signs of failed neuroregeneration than electrodiagnostic procedures and thus ascertain the need of surgical revision (see also Chap. 3) [24].

2.4.2 MRI

2.4.2.1 Background and Overview

Traditionally, MRI has been included in the diagnostic assessment of peripheral neuropathies only to rule out any suspicion of a causative mass lesion that might lead to a compression injury of the adjacent nerve. In the early 1990s, the group of Filler and colleagues started to develop MRI sequences for imaging of the peripheral nervous system in experimental studies which required a higher structural resolution and an increased nerve lesion contrast [12]. They were also the first who termed this optimized MRI technique as magnetic resonance neurography (MRN) [12]. Very early it became clear that heavily T2-weighted (T2w) sequences with fat saturation are most suitable to distinguish between healthy and impaired nerve tissue [3, 22]. While normal nerves appear isointense or slightly hyperintense to muscle tissue in these sequences, the MRN correlate of a nerve lesion is a markedly increased T2w signal or rather an increase in the T2 relaxation time of the corresponding nerve, a marker with high sensitivity but low specificity; for instance, it cannot differentiate between mechanical and immune-mediated, metabolic, or hereditary nerve injury [22]. The spatial extension of the signal change has been first described in experimental animal studies in neurotmetic and axonotmetic nerve injuries. A significant increase of the intraneural T2w signal has been observed in animals to occur within 24–48 h after traumatic axonal damage, not just at the lesion site but also along the distal course of the nerve due to Wallerian degeneration. Nerve regeneration after successful neurosurgical restoration was accompanied by a normalization of the formerly increased T2w signal with a proximo-to-distal course over several weeks [4]. These MRN findings correlated well with both electrophysiological and histological data. In contrast to traumatic nerve injuries with complete or partial nerve discontinuity, the T2w signal increase in demyelinating neuropathies has been found to be restricted to the lesion site without any distal or proximal extension [4].

These first experimental studies provided the basis for the implementation of MRN in the diagnostic workup of peripheral nerve disorders in human patients. Nowadays, it is an accepted technique that allows the direct and precise visualization of nerve injury even on a fascicular microstructural level (fascicular imaging) [28, 34]. That means that MRN can clearly differentiate between nerve lesions affecting the complete cross section of a nerve and partial and therewith fascicular lesions containing often somatotopic information [28]. In traumatic nerve injury, it can visualize the affected individual fascicles, the proximal lesion border, potential distal discontinuity, or nerve compressing masses such as hematoma or bone fragments [34].

2.4.2.2 Technical Requirements

The term MRN implies the application of certain MR pulse sequences that can visualize peripheral nerves and distinguish them from surrounding soft tissue and vessels [12]. The basic requirements to achieve a structural resolution sufficient for imaging of nerve tissue, or rather visualization of nerve fascicular structure, are a high magnetic field strength of 3 Tesla and heavily T2w, fat-saturated sequences with an in-plane resolution of 0.1–0.4 × 0.1–0.4 mm, and a slice thickness of not more than 2–3.5 mm [29]. Fat saturation is crucial to reliably differentiate between bright nerve signal and surrounding fat tissue and can be achieved by either frequency-selective saturation of the fat signal in T2w fast spin echo (SE) sequences or with nulling of the fat signal as it is done in short inversion recovery (STIR) or turbo inversion recovery magnitude (TIRM) sequences [3]. Unenhanced T1w sequences can be beneficial in regions of difficult anatomical orientation, e.g., peripheral nerves emerging from the lumbosacral plexus. Additional application of a contrast agent and subsequent acquisition of T1w sequences with fat saturation are needed in cases of mass lesions like nerve or nerve sheath tumors, but also in remaining or recurring neuropathy after surgical interventions to rule out an over-production of potentially nerve compromising scar tissue [22].

2.4.2.3 MRN Findings in Nerve Injury

In patients with traumatic brachial plexus injuries, it is of utmost importance to early differentiate between a nerve root avulsion from the spinal cord, also referred to as a preganglionic lesion, and a postganglionic nerve lesion, involving the supra- or infraclavicular parts of the brachial plexus (trunks, divisions, cords, branches) [32]. A total nerve root avulsion can be easily diagnosed with conventional imaging methods, such as CT myelography or spinal MRI, which will show an unencapsulated pouch of fluid due to the extravasation of cerebrospinal fluid (CSF), commonly termed pseudomeningocele (Fig. 2.3). Most complete tears or avulsions of nerve roots or ganglia cannot be grafted due to the retraction of proximal nerve

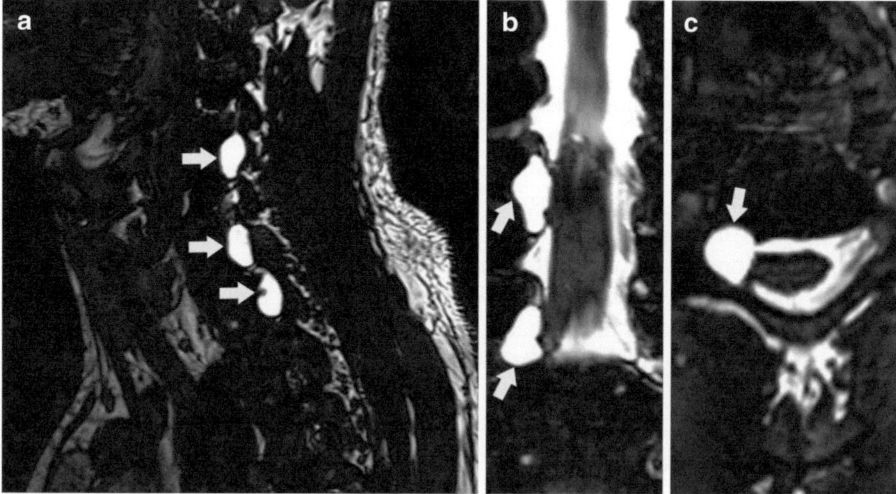

Fig. 2.3 MRN: 3D constructive interference in steady state (CISS) sequence in sagittal (**a**), coronal (**b**), and transversal (**c**) reformations. *Arrows* point to pseudomeningoceles of the C6, C8, and Th 1 nerve root, representing complete traumatic nerve root avulsions

tissue. In case of postganglionic brachial plexus lesions, a further differentiation between a complete separation of proximal and distal nerve ends (neurotmesis) without any chance of spontaneous recovery (Fig. 2.4) and an incomplete or partial nerve discontinuity or stretching injury (neuropraxia or axonotmesis), which might recover without surgical treatment, is essential for an adequate surgical planning and also for a prognostic estimation. Besides direct visualization of nerve discontinuity (Fig. 2.4), the most obvious MRN sign of a complete nerve transsection is an end-bulb neuroma (EBN; Fig. 2.5a), while an incomplete nerve lesion might show

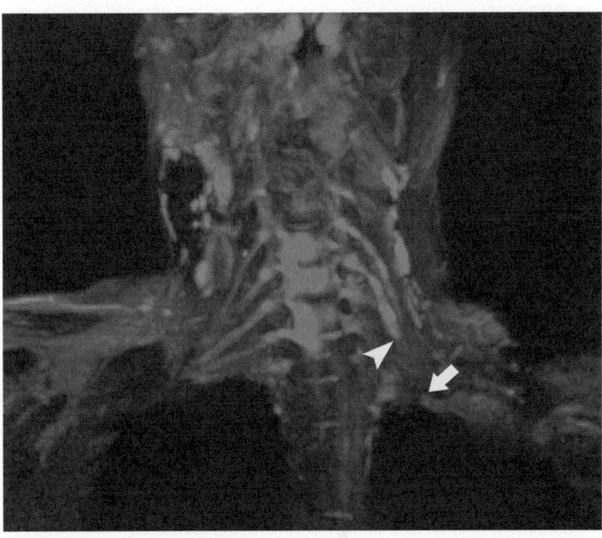

Fig. 2.4 MRN: 3D T2-weighted inversion recovery sampling perfection with application optimized contrasts using different flip angle evolution (SPACE) sequence. Coronal reconstruction. Postganglionic brachial plexus lesions showing neurotmesis. *Arrowhead* points to the proximal, and *arrow* points to the distal nerve ends. Note the remarkable distance between proximal and distal nerve ends caused by retraction

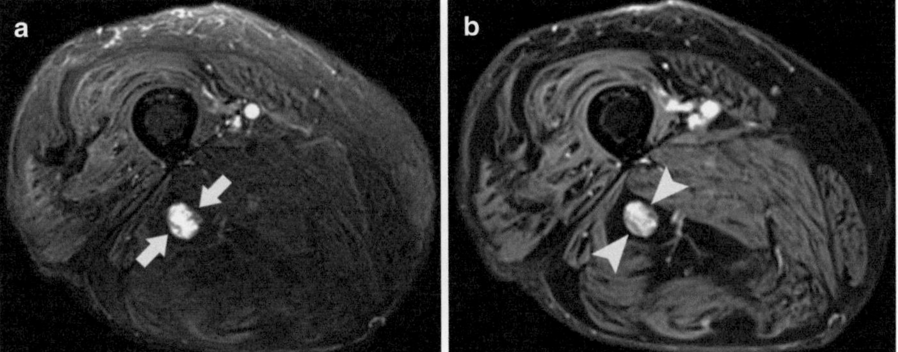

Fig. 2.5 MRN: axial T2-weighted sequence with spectral fat saturation (**a**) and axial contrast-enhanced T1-weighted sequence with spectral fat saturation (**b**). Note the markedly increased cross-sectional diameter and intraneural T2w signal (*arrows*) of the sciatic nerve at mid-thigh level representing an end-bulb neuroma after traumatic amputation of the right leg at knee level. After the application of a contrast agent, the end-bulb neuroma shows the typical increased, slightly inhomogeneous enhancement (*arrowheads*)

a neuroma-in-continuity (NIC) [14, 23]. A neuroma consists of fibroneural tissue that develops as a result of failed nerve regeneration with a multidirectional proliferation of cells as well as distortion of normal nerve architecture. Both EBN and NIC show continuity with the proximal parent nerve, but only NIC shows continuity with the distal parent nerve as well [1]. Typically, traumatic neuromas present not only with an increased, often heterogeneous, T2w signal but also with a marked increase of the cross-sectional diameter and an oval or nodular form on coronal or sagittal images (Fig. 2.5) [33]. After the application of a contrast agent, it will show an increase and also a heterogeneous enhancement (Fig. 2.5b) [1, 32]. Even in the absence of a defined neuroma, an increase of the nerve diameter might be visible, which is another, this time morphometric and therewith signal independent, MRN sign of nerve damage. It is not just related to neuromas but is an unspecific MRN sign of nerve impairment in general.

The average speed of nerve regrowth is around 1 mm per day, so that a complete recovery can take months or even years. During that time, it might be difficult to monitor proper nerve regeneration and failure of axonal regrowth, e.g., due to dislocation of the proximal and distal nerve ends or fascicles, which would require early surgical therapy (see also Chap. 3). MRN with its direct visualization of nerve lesions can help to monitor the physiological regrowing in that it shows the disappearance of the formerly bright T2w signal with a proximo-to-distal gradient [10].

Any interpretation of an increased nerve T2w signal has to be made carefully and with the knowledge that the T2w signal might be artificially increased related to certain specific MRI artifacts like the magic angle effect. In the majority of cases, it can be avoided easily by positioning the examined extremity with an alignment of less than 30° relative to the B0 field direction [18]. However, in examinations of the cervical and lumbar plexus, this alignment is not always realistic due to the normal anatomical course of the emerging nerves, and it is important that these angulation-induced signal changes are not mistaken for true pathologic nerve lesions [3].

Determining the existence or absence of physiological nerve repair is crucial and might be diagnostically challenging even when changes in the proximo-to-distal extend of nerve T2w signal all together with clinical and electrophysiological examinations are performed (see also Chap. 3). Current studies discuss the validity and diagnostic advantage of new techniques, such as diffusion tensor tractography (DTT), a method based on direction-dependent diffusion in anisotropic structures or rather anisotropic movement of water molecules, to monitor early nerve regeneration in vivo. First results of animal studies showed that tracked fibers terminate at the point of axonal discontinuity within hours after traumatic nerve injury, while fibers may extend distal to the located injury and show an increase in the fractional anisotropy (FA) in case of nerve regeneration [25, 35]. However, further investigations and correlation with clinical and electrophysiological measurements are needed to estimate the diagnostic benefit and outcome.

The most frequent traumatic injuries of the sciatic nerve are iatrogenic and are either induced by gluteal injection injury or periprocedural in hip replacement surgery. In case of iatrogenic trauma related to hip replacement surgery, direct imaging identification of the exact lesion site and determination of injury severity are often challenging, due to metal artifacts that are related to metal implants in the direct vicinity of the nerve. However, new techniques of artifact reduction make it still possible to achieve a nerve lesion contrast that is sufficient for precise lesion localization [39]. Besides complex adjustments of sequence parameters whose description would exceed the purpose of this book, the following aspects should be kept in mind: spin echo (SE) or turbo spin echo (TSE) sequences are more beneficial than gradient echo (GRE) sequences, as the 180° refocusing pulse used in SE sequences corrects for large magnetic field inhomogeneities and therewith reduces dephasing artifacts. Additionally, STIR sequences should be used for the necessary fat suppression in T2w sequences, as they are less dependent on a homogenous magnetic field than spectral fat saturation techniques [39]. Furthermore, the acquisition of T1w sequences might be beneficial as they are less vulnerable to susceptibility artifacts while providing sufficient anatomical resolution for the detection of nerve discontinuity or compromising material [39].

Clinical and electrophysiological measurement at proximal sites of the lower extremities often lacks to precisely localize the nerve lesion as well as to give an estimation of the regenerative potential without surgical therapy. MRN has been proven to be able to directly visualize the exact lesion site with high sensitivity by evaluating the typical MRN pattern as described before, but can also give a detailed pathomorphological description of the injury, like overproduction of epi- or intraneural scar tissue, development of neuroma, or extent of fascicular involvement [30].

2.4.2.4 MRN Findings in Denervated Muscle

Not only the nerves but also the muscle tissue is visualized through MRN. Normal muscle tissue presents with an intermediate signal on T1w and T2w sequences. In acute nerve injury and subsequent denervation of dependent muscles, a marked increase of the T2 relaxation time can be observed as early as 5 days after an axonotmesis or neurotmesis [38]. MRI findings well correlate with the amount of spontaneous activity on EMG [16, 26] and with the size of the MUAPs [16]. Its main advantages, namely, its painlessness and its ability to visualize the whole cross section of an extremity (Fig. 2.6), must be balanced against its lower sensitivity to axonotmesis and its blindness to neurapraxia [26]. Overall, in certain clinical situations, it may be particularly helpful to complement electrodiagnostic procedures and MRN.

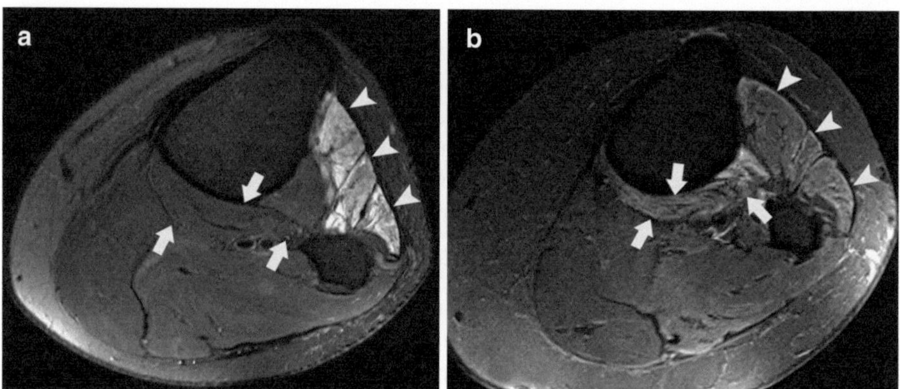

Fig. 2.6 MRN: axial T2-weighted sequence with spectral fat saturation (**a**, **b**). A lesion of the common peroneal nerve (**a**) leads to an increased T2w signal of depending muscles (*arrowheads* point to the anterior tibial, extensor longus, and long peroneal muscles) which is the MRN correlate of muscle denervation. In contrast to that, the pattern of an L5 radiculopathy (**b**) shows an additional denervation signal nerve in the posterior tibial and popliteus muscle (*arrows* in **b**; *arrows* in **a** show the normal signal intensity of the same muscles)

References

1. Ahlawat S, Belzberg AJ, Montgomery EA, et al. MRI features of peripheral traumatic neuromas. Eur Radiol. 2016;26(4):1204–12.
2. Angerer M, Kleudgen S, Grigo B, et al. Hochauflösende Sonographic des Nervus medianus – neue sonomorphologische Untersuchungstechnik peripherer Nerven. Klin Neurophysiol. 2000;31(02):53–8.
3. Bendszus M, Stoll G. Technology insight: visualizing peripheral nerve injury using MRI. Nat Clin Pract Neurol. 2005;1(1):45–53.
4. Bendszus M, Wessig C, Solymosi L, et al. MRI of peripheral nerve degeneration and regeneration: correlation with electrophysiology and histology. Exp Neurol. 2004;188(1):171–7.
5. Bischoff C, Pöschl P, Dreger J, et al. Kontra – Bilduntersuchungen in der Diagnostik der Erkrankungen peripherer Nerven – nur in speziellen Ausnahmen sinnvoll und notwendig. Klin Neurophysiol. 2015;46(02):93–6.
6. Bischoff C, Schulte-Mattler W. Das EMG-Buch. 4th ed. Stuttgart: Thieme; 2015.
7. Böhm J, Schelle T. Stellenwert der hochauflösenden Sonografie bei der Diagnostik peripherer Nervenerkrankungen. Akt Neurol. 2013;40(05):258–68.
8. Campbell WW. Evaluation and management of peripheral nerve injury. Clin Neurophysiol. 2008;119(9):1951–65.
9. Coraci D, Pazzaglia C, Doneddu PE, et al. Post-traumatic neuroma due to closed nerve injury. Is recovery after peripheral nerve trauma related to ultrasonographic neuroma size? Clin Neurol Neurosurg. 2015;139:314–8.
10. Dailey AT, Tsuruda JS, Filler AG, et al. Magnetic resonance neurography of peripheral nerve degeneration and regeneration. Lancet. 1997;350(9086):1221–2.
11. Eder M, Schulte-Mattler W, Pöschl P. Neurographic course of Wallerian Degeneration after human peripheral nerve injury. Muscle Nerve. 2016. doi: 10.1002/mus.25489. [Epub ahead of print].
12. Filler AG, Howe FA, Hayes CE, et al. Magnetic resonance neurography. Lancet. 1993; 341(8846):659–61.
13. Fornage BD. Peripheral nerves of the extremities: imaging with US. Radiology. 1988;167(1): 179–82.

14. Grant GA, Britz GW, Goodkin R, et al. The utility of magnetic resonance imaging in evaluating peripheral nerve disorders. Muscle Nerve. 2002;25(3):314–31.
15. Hoffmann P, Buck-Gramcko D, Lubahn JD. The Hoffmann-Tinel sign. 1915. J Hand Surg Br. 1993;18(6):800–5.
16. Jonas D, Conrad B, von Einsiedel HG, et al. Correlation between quantitative EMG and muscle MRI in patients with axonal neuropathy. Muscle Nerve. 2000;23(8):1265–9.
17. Jürgens TP, Puchner C, Schulte-Mattler WJ. Discharge rates in electromyography distinguish early between peripheral and central paresis. Muscle Nerve. 2012;46(4):591–3.
18. Kastel T, Heiland S, Baumer P, et al. Magic angle effect: a relevant artifact in MR neurography at 3 T? AJNR Am J Neuroradiol. 2011;32(5):821–7.
19. Kimura J. Electrodiagnosis in diseases of nerve and muscle. 4th ed. Oxford: Oxford University Press; 2013.
20. Koenig RW, Pedro MT, Heinen CP, et al. High-resolution ultrasonography in evaluating peripheral nerve entrapment and trauma. Neurosurg Focus. 2009;26(2):E13.
21. Koenig RW, Schmidt TE, Heinen CP, et al. Intraoperative high-resolution ultrasound: a new technique in the management of peripheral nerve disorders. J Neurosurg. 2011;114(2):514–21.
22. Kollmer J, Bendszus M, Pham M. MR neurography: diagnostic imaging in the PNS. Clin Neuroradiol. 2015;25(Suppl 2):283–9.
23. Kuntz C, 4th, Blake L, Britz G et al. Magnetic resonance neurography of peripheral nerve lesions in the lower extremity. Neurosurgery. 1996;39(4):750–6; discussion 756–7.
24. Lauretti L, D'Alessandris QG, Granata G et al. Ultrasound evaluation in traumatic peripheral nerve lesions: from diagnosis to surgical planning and follow-up. Acta Neurochir. 2015; 157(11):1947–51; discussion 1951.
25. Lehmann HC, Zhang J, Mori S, et al. Diffusion tensor imaging to assess axonal regeneration in peripheral nerves. Exp Neurol. 2010;223(1):238–44.
26. McDonald CM, Carter GT, Fritz RC, et al. Magnetic resonance imaging of denervated muscle: comparison to electromyography. Muscle Nerve. 2000;23(9):1431–4.
27. Padua L, Di Pasquale A, Liotta G, et al. Ultrasound as a useful tool in the diagnosis and management of traumatic nerve lesions. Clin Neurophysiol. 2013;124(6):1237–43.
28. Pham M, Baumer P, Meinck H-M, et al. Anterior interosseous nerve syndrome: fascicular motor lesions of median nerve trunk. Neurology. 2014;82(7):598–606.
29. Pham M, Bäumer T, Bendszus M. Peripheral nerves and plexus: imaging by MR-neurography and high-resolution ultrasound. Curr Opin Neurol. 2014;27(4):370–9.
30. Pham M, Wessig C, Brinkhoff J, et al. MR neurography of sciatic nerve injection injury. J Neurol. 2011;258(6):1120–5.
31. Schulte-Mattler WJ, Georgiadis D, Tietze K, et al. Relation between maximum discharge rates on electromyography and motor unit number estimates. Muscle Nerve. 2000;23(2):231–8.
32. Silbermann-Hoffman O, Teboul F. Post-traumatic brachial plexus MRI in practice. Diagn Interv Imaging. 2013;94(10):925–43.
33. Singson RD, Feldman F, Staron R, et al. MRI of postamputation neuromas. Skeletal Radiol. 1990;19(4):259–62.
34. Stoll G, Bendszus M, Perez J, et al. Magnetic resonance imaging of the peripheral nervous system. J Neurol. 2009;256(7):1043–51.
35. Takagi T, Nakamura M, Yamada M, et al. Visualization of peripheral nerve degeneration and regeneration: monitoring with diffusion tensor tractography. Neuroimage. 2009;44(3):884–92.
36. Tinel J. Le signe du "fourmillement" dans les lésions des nerfs périphériques. Presse Med. 1915;23:388–9.
37. Toros T, Karabay N, Ozaksar K, et al. Evaluation of peripheral nerves of the upper limb with ultrasonography: a comparison of ultrasonographic examination and the intra-operative findings. J Bone Joint Surg Br. 2009;91(6):762–5.
38. West G A, Haynor D R, Goodkin R et al. Magnetic resonance imaging signal changes in denervated muscles after peripheral nerve injury. Neurosurgery. 1994;35(6):1077–85; discussion 1085–6.
39. Wolf M, Baumer P, Pedro M, et al. Sciatic nerve injury related to hip replacement surgery: imaging detection by MR neurography despite susceptibility artifacts. PLoS One. 2014; 9(2):e89154.

Timing and Decision-Making in Peripheral Nerve Trauma

3

Hans Assmus

Contents

3.1 Introduction

The decision processes during diagnosis and treatment of an injury of the peripheral nerve are much more complex than, for example, those for injuries of bones or tendons. The nature and cause of the injury, its localization, and its depth/severity require very distinct decisions with regard to timing and technique for intervention. Timing of the nerve repair is important.

Basically, a sharply and neatly transected nerve is to be judged different than a nerve that has been bluntly transected or violently torn. While the first is usually caused by a cut, the latter two occur during injuries with bony fractures, gunshots, or electrical and other physical traumata.

H. Assmus
Abtweg 13, D-69198 Schriesheim, Germany
e-mail: hans-assmus@t-online.de

© Springer International Publishing AG 2017
K. Haastert-Talini et al. (eds.), *Modern Concepts of Peripheral Nerve Repair*,
DOI 10.1007/978-3-319-52319-4_3

A sharp transection injury within a clean (not contaminated) wound requires immediate repair, while in the majority of cases, a wait-and-see attitude and an (early) secondary nerve reconstruction is advisable. For appropriate decision-making, a good knowledge of the pathophysiological conditions in the peripheral nervous system and of related processes, which occur at the same time in the central nervous system, is helpful. The crucial role of the time factor in nerve reconstruction relates to the early induction of the regeneration processes by intracellular signaling of the axotomized neurons and their neighboring nonneural cells [9].

There is no debate that nerve reconstruction performed as early as possible strongly increases the prognosis and condition for optimized functional recovery. The quality of motor recovery is continuously decreasing 6 months after injury [5, 6, 15]. This limitation is less strong for sensory recovery. It has to be avoided, however, to disrupt a possibly ongoing spontaneous recovery after a completely reversible nerve block (neurapraxia, see Chap. 1). Overhasty, but also too broad resection of a neuroma in continuity after gunshot, contusion, or stretching or of a neuroma after partial transection could result in poorer outcome than an observant approach. Modern techniques like electrodiagnostical nerve stimulation or neurosonography considerably facilitate the diagnostic classification of peripheral nerve injuries and result into early nerve reconstruction where indicated.

3.2 The Decision Process

A general outline of the complex decision process for nerve reconstruction surgery and its timing is illustrated in Fig. 3.1. Depending on whether an open or closed injury exists, it has to be decided for a primary or secondary reconstruction surgery. During the decision process, the wound condition, the depth of the injury, and eventually also results of an exploratory exposure of the nerve or from imaging or electrophysiological evaluations have to be considered. During primary or early secondary reconstruction, additional decision has to be taken with regard to the appropriate reconstruction technique. During a regular postoperative monitoring, the regeneration process has to be documented, and eventually new diagnostic and surgical decisions have to be taken. A mean monitoring and treatment period covers approximately 2 years and more.

The complex challenge of nerve repair and reconstruction, including additional steps that may become necessary, is exemplarily outlined by the practical case example below. Subject is a typical and frequent lower arm injury with disconnection of the ulnar nerve.

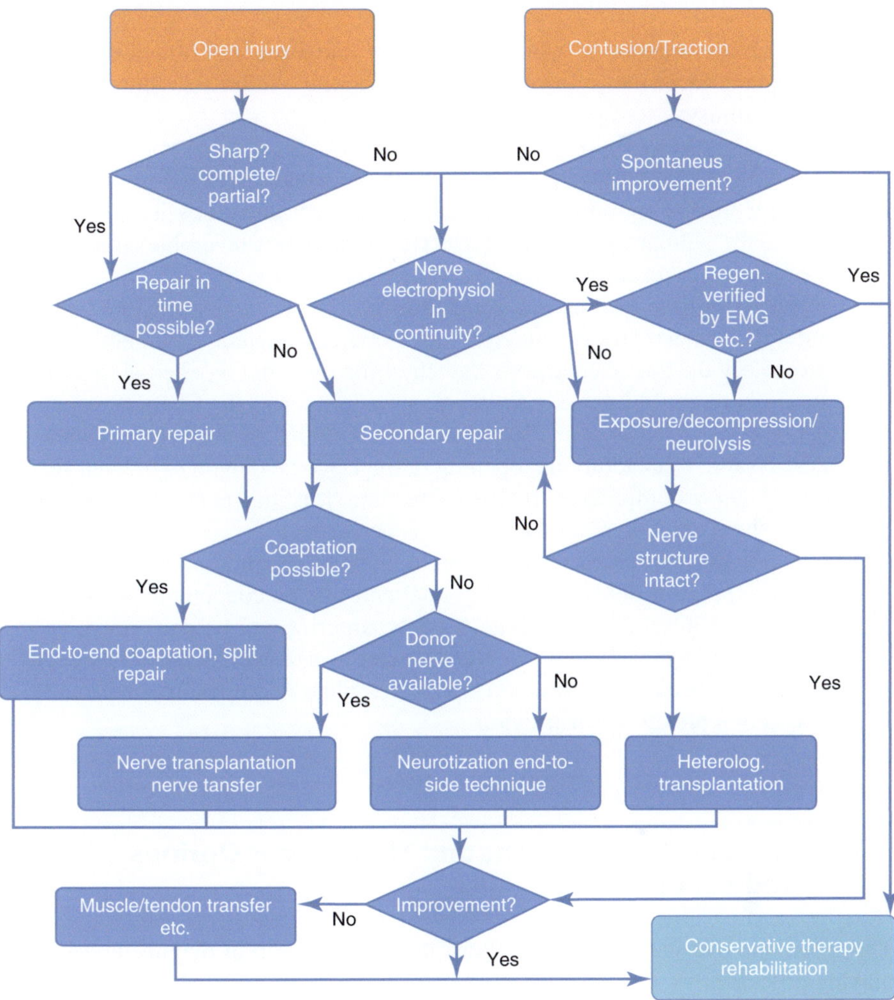

Fig. 3.1 Algorithm of the diagnostic and therapeutic steps and treatment strategies for nerve injury

Case Example

A younger patient sustained a laceration of the left volar proximal forearm 1 month ago with transection of the ulnar nerve. During primary wound care, the nerve ends had been loosely approximated.

The nerve surgeon, to whom the patient had been referred for further management, faces the following options depending on the assessment of the pattern of injury:

- Attempt of an end-to-end coaptation of the ulnar nerve. Therefore, the nerve would be mobilized over a long distance followed by its transposition to the palmar side of the elbow. This approach is favorable when only a small defect resulted from a sharp transection of the nerve.
- If a tension-free end-to-end coaptation is not possible, decision for an autologous nerve transplantation has to be considered. As donor nerve preferably the dorsal cutaneous branch of the ulnar nerve, alternatively the sural nerve could be used. Artificial nerve conduits (see Chap. 8) cannot yet be considered as general alternatives to autologous nerve transplants.
- Discussable is an additional opening of the Loge de Guyon to prevent secondary nerve compression by tissue swelling. It is indispensable, however, when the technique mentioned below is performed.
- Not generally accepted is the option to increase the motor function of the intrinsic hand muscles by an additional end-to-side coaptation (see Chap. 4) between the final motor branch of the anterior interosseous nerve and the deep branch of the ulnar nerve. The end-to-side technique is, however, more often used in repair of very proximal ulnar nerve lesions, especially in brachial plexus reconstruction.

3.3 Preoperative Decisions and Therapeutic Options (Indication)

The schematic overview in Fig. 3.2 illustrates the lining up of diagnostic decision-making processes.

After a trauma with assumed lesion of a peripheral nerve, it applies first to prove or to exclude the same. In the case of a proven nerve injury, its depth has to be documented. Careful anamnesis could often point on a nerve injury, e.g., a report of violent shooting in electrifying pain in the context of a surgical intervention. In the case of a traumatic accident, its course and primary clinical status are to be exactly documented, not only for insurance-legal reasons but also to answer the question whether the lesion has developed by the accident mechanism (e.g., avulsion injury) or the supply of a fracture (e.g., with a humerus fracture).

The severity level of the nerve lesion has to be classified as this is indispensable for the statement of the injury depth and the prognosis (see Chap. 1) but serves equally for the decision-making regarding the treatment strategy, in particular the indication for the operation. The Medical Research Council (MRC) scale (http://www.mrc.ac.uk/research/facilities-and-resources-for-researchers/mrc-scales/mrc-muscle-scale/) and the Seddon and Sunderland classifications (with the

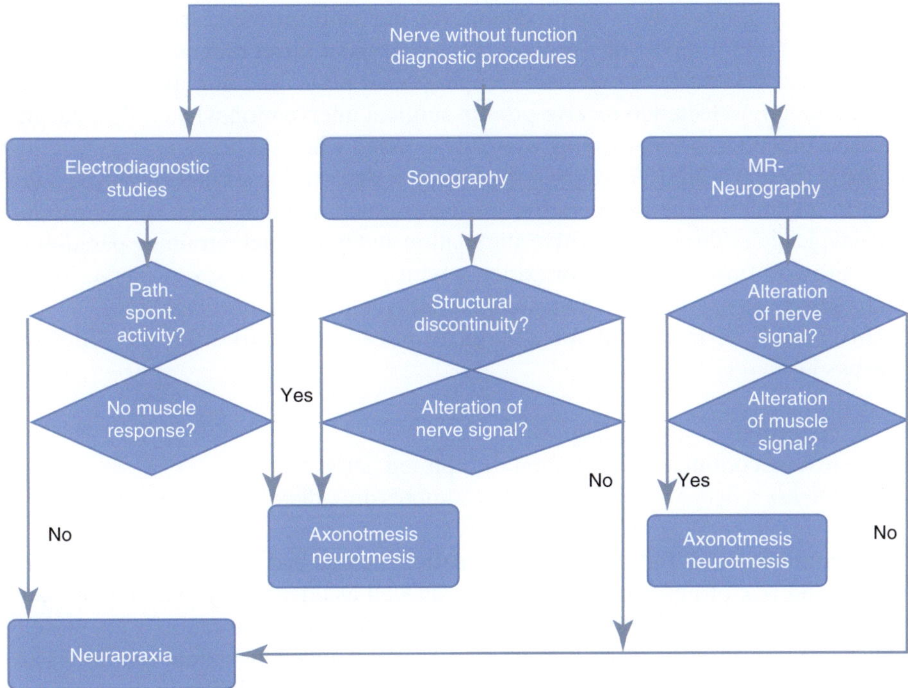

Fig. 3.2 Primary diagnostic procedures (see also Chap. 2)

additions of Millesi) are still commonly used (see Chap. 1). Especially in early stages after nerve lesion, it can be difficult or impossible to differentiate between neurapraxia, axonotmesis, and neurotmesis just on the basis of clinical symptoms and physical findings. Therefore, additional electrophysiological investigations are essential [17]. In early stages after a nerve injury, compound muscle action potentials (CMAPs) and motor units (MU) can be examined, which are, however, of only limited evidence. Recently, nerve sonography showed its superiority over the electrophysiological assessments [7]. But also nerve sonography has its limitations, as it is very much restricted to the examination of rather superficial nerves. Deep-running nerves can be judged, however, with the magnetic resonance neurography (MRN) (see Chap. 2). Already today, this modern method substantially affects diagnostic and therapeutic decision-making processes and in particular also the planning of the surgical intervention [8]. Ultrasound or standard MRN are unable to fully discriminate between neurotmesis and axonotmesis, in particular when the nerve remains in continuity. Diffusion tensor tractography (DTT) represents a recent development in MR imaging that may revolutionize this aspect of the diagnosis and monitoring of peripheral nerve trauma [19].

Sunderland Grade III lesions may often regenerate spontaneously (incomplete regeneration) and then result in better functional recovery than after nerve reconstruction. Both will result in incomplete regeneration, but the degree may vary considerably. Especially in this scenario, the decision for the correct procedure is very important for the further process. On the basis of novel diagnostic techniques, in particular related to the advances in peripheral nerve imaging, patients may be more correctly selected to receive prompt surgical intervention when indicated, and invasive procedures may be more correctly avoided when spontaneous regeneration is likely. Simon et al. [20] suggest that "further exploration of non-invasive strategies to augment nerve regeneration processes, such as modulation of central and axonal plasticity through repetitive stimulation and functional retraining paradigms, may provide further benefit for patients with moderate and severe nerve injury, including those patients in whom surgical intervention is not needed" [20].

In general, an indication for a surgical procedure results from the following reasons [4]:

- To confirm or establish diagnosis
- To restore continuity to a severed or ruptured nerve
- To release a nerve of an agent that is compressing, distorting, or occupying it

Contraindications for a surgical intervention include bad general condition of the patient and risk of general or local sepsis, as well as uncertainty over the kind and extent of the injury (e.g., bullet or saw injury). A nerve reconstruction is not reasonable whenever no appropriate instrumental, machine-aided, or spatial conditions are present and whenever no experienced operation team is available. In certain cases also a primary muscle or tendon transfer can be the better choice, e.g., in the case of irreparable lesions of predominantly motor nerves (such as of the radial and common peroneal nerves).

In fact, the ideal case of a primary treatment of a nerve injury, i.e., the smooth uncomplicated disconnection of a nerve, is rather the exception. For the far more frequent secondary treatment, temporal limits have been specified, which vary depending upon the examiner (Fig. 3.3).

Acute peripheral nerve lesions require a differentiated proceeding. The strategic decision for an immediate reconstruction or a wait-and-see attitude depends on the type and the depths of the lesion. Generally a complete/total nerve transection injury can be assumed whenever an open cut or stab injury exists along the nerve trajectory and the loss of the nerve function has been proven clinically. During wound treatment the discontinuity of the nerve will be confirmed, and under optimal conditions, primary nerve reconstruction will be performed. This is not only the simplest decision to be taken by the nerve surgeon but also the ideal scenario for the patient with the best prognosis. Under ideal conditions the prerequisites are as follows: (1) smooth sharp transection and clean wound properties, (2) experienced nerve surgeon, and (3) appropriate technical equipment and surgery room. Certain types of nerve injury and their extent together with the extent of accompanying injuries and other concomitant factors, however, do often prohibit primary nerve repair.

Most examiners do not define a strict time limit for the primary coaptation. The transition to an early secondary treatment runs smooth, although an intervention within the first week ($\leq$ 10 days) after injury is generally referred to as the optimum.

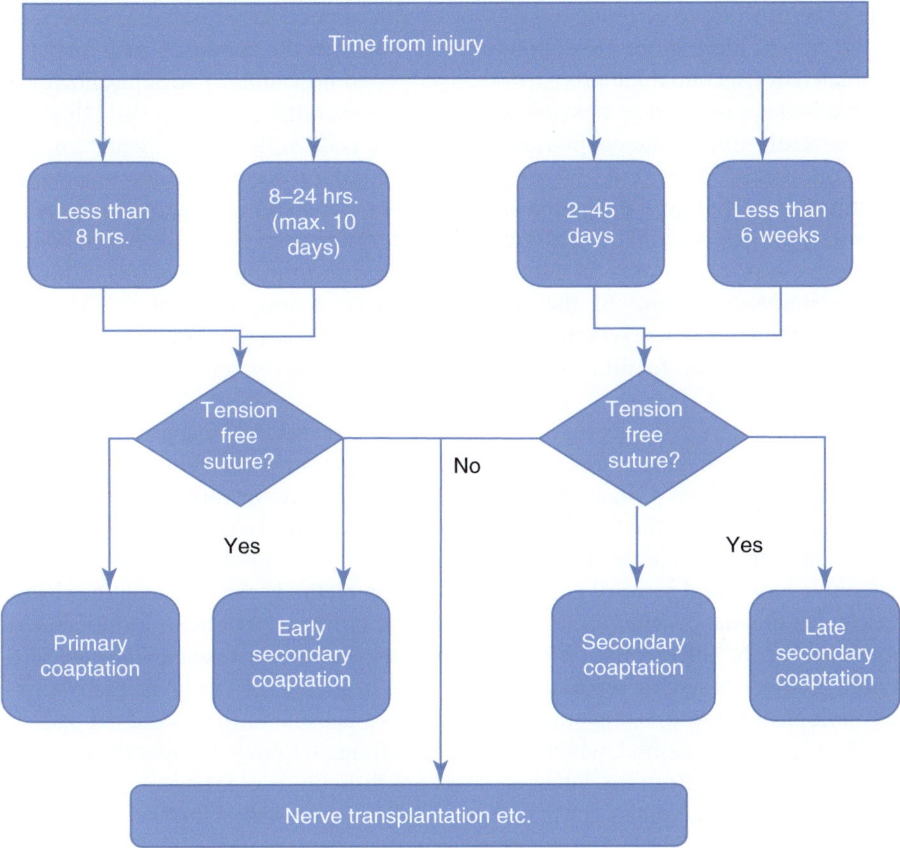

Fig. 3.3 Timing and temporal limits of primary and secondary peripheral nerve repair

A maximum time window of 6 weeks is accepted for secondary nerve repair. Such secondary nerve repair may not always have an inferior prognosis, because a highly experienced nerve surgeon performing a meticulous and conservative approach may compensate the disadvantages of the delayed treatment.

The decision to accomplish a reconstruction after more than 6 months must be discussed with the patient and depends on many factors like the age of the patient, the kind of nerve injured (motor or sensory), the proximo-distal height of the injury, concomitant lesions, etc.

3.4 Intraoperative Decisions

The main principle is to perform tension-free and optimal adaptation of the nerve ends and fascicles. Contused nerve stumps have to be trimmed back to the healthy epineurium and a visible fascicular structure. During an immediate repair approach, it may be difficult to estimate exactly the length of necessary resection, especially in the case of more blunt dissection injuries. This specific condition may be better evaluated in a secondary approach or delayed repair procedure. The delay, however, should be kept as short as possible since nerve ends retract with time and this will consequently impair the coaptation of the nerve ends without significant tension. For combined lesions (cut and crush), decision-taking for immediate or delayed treatment is particularly difficult and responsible. The latter is also true for gunshot injuries in which typically a massive, diffuse destruction of tissue is present. Once, in these cases, a neuroma in continuity is found during the secondary care, the intraoperative recording of the compound nerve action potential (cNAP) can deliver reliable evidence whether nerve conduction at the lesion site is preserved or not. Such evidence facilitates very much the decision whether to remove the neuroma or not [13].

After the decision for nerve reconstruction has been taken, the appropriate reconstruction approach has to be considered. The intraoperative decision-taking process is illustrated in Fig. 3.4. In addition further important intraoperative determinations have to be made, like the length of the resection of the nerve stumps (how much to resect) and the anatomical allocation of the nerve bundle groups. The use of intraoperative motor and sensory nerve differentiation methods can diminish the risk of fascicular mismatch when grafting an autologous nerve. Available intraoperative methods for the differentiation between sensory and motor fascicles are the anatomic method, based on separate identification of groups of fascicles, the electrophysiological method, using stimulation of nerve fascicle in the awakened patient, and histochemical method, which is based on staining for enzymes specific to motor or sensory nerves. Both the latter methods are difficult to use, because the patient has to be highly compliant for the electrodiagnostic evaluation and because of the time gap that develops before sample tissue has been analyzed histochemically.

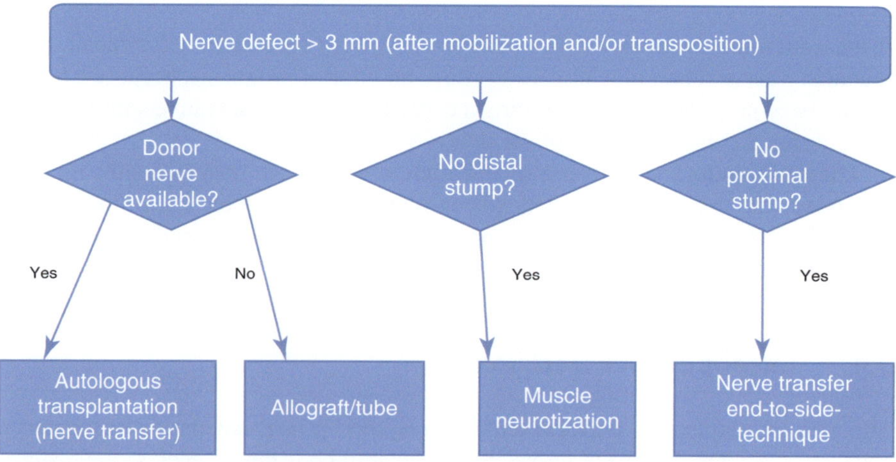

Fig. 3.4 Decision-taking depending on the length of the nerve defect

Finally, the appropriate donor nerve has to be identified. Most of the time, the sural nerve is selected, but the cutaneous antebrachial and the saphenous nerve comprise good alternatives.

Special techniques alternative to autologous nerve grafting and their limitations are described and discussed in the Chaps. 4 and 7.

Harvesting the donor nerve can be done by different methods: for the sural nerve, several small incisions will be made along the nerve course, or a nerve stripper will be used [3], which reduces the operation time significantly and gives the best cosmetic results. The decision depends on personal preferences of the nerve surgeon.

An additional consideration is needed with regard to the degree of justifiable tension at the coaptation site. An epineural coaptation with little tension is the method of choice [21]. This is usually achieved by sufficient mobilization of the nerve as well as by transposition of the nerve (e.g., transposition of the ulnar nerve into the cubital fossa) or through slight joint bending. A rough indication for what degree of tension is barely acceptable at the coaptation site can give the application of 10-0 sutures. Whether fibrin glue is an alternative or additive to suture is discussed in Chap. 4.

Techniques like end-to-side coaptation and the direct muscular neurotization (only a treatment at ultima ratio) are not commonly accepted as alternative approaches for nerve reconstruction (see Chap. 4). The latter techniques may be considered in reconstruction of extensive proximal lesions or extended brachial plexus lesions (see Chap. 6).

3.5 Indication in Case of Partial Lesions and Neuroma Formation

Treatment of neuroma is generally challenging and often results in disappointing results. The decision for or against neuroma resection may be difficult to take and deserves in-depth discussion with the patient. This is also true for neuroma in continuity, e.g., those of the median nerve at the wrist resulting from superficial cut injuries. Such neuroma can only be treated with the so-called split-repair by removing only neuromatous parts and preservation of those nerve fascicles that are still functional. In these cases intraoperative sonography and cNAP studies could be helpful to evaluate intact nerve fascicles (see Chaps. 2 and 5). The performance of a split-repair is challenging, can only be done using a surgery microscope or magnifying glasses, and requires specific expertise from the responsible nerve surgeon.

3.6 Decisions in or Concerning Combined Lesions

Due to the fact that traffic accidents represent the most frequent cause of peripheral nerve injuries, examiners are frequently confronted with combined lesions. Combined lesions significantly impact the timing of nerve repair, because the primary attempt is the treatment of bony injuries, e.g., the primary stabilization of one or more fractures. Treatment of a vascular lesion is additionally prior ranking and could represent the first brick of the treatment chain in case of life-threatening blood loss. It has further to be considered that a tendon suture should not be performed together with a nerve reconstruction because the aftercare of both is different, immobilization versus mobilization, respectively. A radial nerve palsy resulting from a fracture of the humerus bone, again, is confronting the nerve surgeon with the need for decision between early exploration and wait-and-see attitude. What the best strategy can be is still controversially discussed. Since it is generally accepted that the radial nerve palsy spontaneously recovers with a rate of more than 70%, the wait-and-see concept is favored in closed humerus shaft fractures [10].

3.7 Perioperative Techniques: Choice of Anesthesia and Application of Tourniquet

With the decision taken for a nerve reconstruction, the question for the appropriate anesthesia arises. General anesthesia is not required for all approaches. Basically, small nerves travelling superficially can also be treated in local anesthesia. For larger nerves of the extremities, regional anesthesia may be sufficient in some cases; mostly general anesthesia is preferred by the surgeon because surgery under regional anesthesia requests a very high compliance from the patient.

The need of bloodless conditions is controversially discussed as they are obligatory among hand and plastic surgeons, but less common among neurosurgeons. The reason for this is that the procedure was thought to be linked to complications and possible damages. Distal to the tourniquet, the nerve line expires within 15–45 min; however, it very rapidly recovers after opening of the cuff, e.g., within a few minutes [14]. The technique can be used without any risks whenever properly performed (broad, well-padded cuff, a pressure of 50–75 mm Hg below systolic pressure, for a maximum of 1.5–2 h). In case bloodless conditions are needed for an extended time period, renewed application of the cuff is possible after transient opening. During the nerve transplantation procedure, however, the tourniquet is opened. A sterile cuff should be used whenever it needs to be placed close to the operating field. The advantage of the bloodless conditions is enabling of a rapid and careful preparation under very good overview of the operation field. Furthermore, small nerve branches can more easily be preserved and small vessels punctually coagulated.

3.8 Postoperative Decisions and Revision Surgery

The proper timing of nerve repair approaches plays an important role not only at the beginning of the treatment but also after nerve reconstruction has been performed. Right after the intervention, it remains uncertain whether it will result in successful functional recovery or not. The regeneration process is to be traced carefully and by means of follow-up examinations in 6-week intervals. While protection sensitivity usually returns spontaneously, tactile discrimination ability returns spontaneously in only a limited number of cases. To support recovery of the latter, it requires purposeful sensory reintegration programs [18]. Therefore, the stimulation of the relearning process must begin immediately after the nerve reconstruction has been performed [9]. Two promising adjunct procedures exist that have been evaluated intensively in animal studies and are about to be translated into clinical application. This is 1 h low-frequency electrical stimulation of the proximal nerve end at the time of reconstruction and daily rehabilitation exercise [12]. A general suggestion to apply the procedures, however, does not yet exist.

Whenever recovery of sensitivity and motor function is missing after reasonable time from primary or secondary nerve reconstruction, the puzzling question of a surgical revision arises. In average a 6-month follow-up period should be considered, and not only the condition of the nerve itself but also the patient's individual conditions (age of the injury and the patient, height/localization of the lesion, eventual concomitant lesions) are of concern. As illustrated in Fig. 3.5, while evaluating the nerve condition, the nerve surgeon does not exclusively depend on a clinical process observation performed with the utmost care, but needs to additionally consider results from electrodiagnostic and imaging assessments. In early stages of reinnervation, it is not possible to record a sensory (or motor) nerve/muscle action potential. Sensory nerve function, however, can be evaluated also in early reinnervation stages by recording of somatosensory evoked potentials (SSEPs) [2]. Weak peripheral volleys within the SSEP may emerge, and a central magnification can enable detection of a cerebral response also in cases where no peripheral NAP can be recorded.

With regard to modern imaging techniques, diffusion tensor tractography (DTT), a special technique of MR neurography, is described to be promising in visualizing

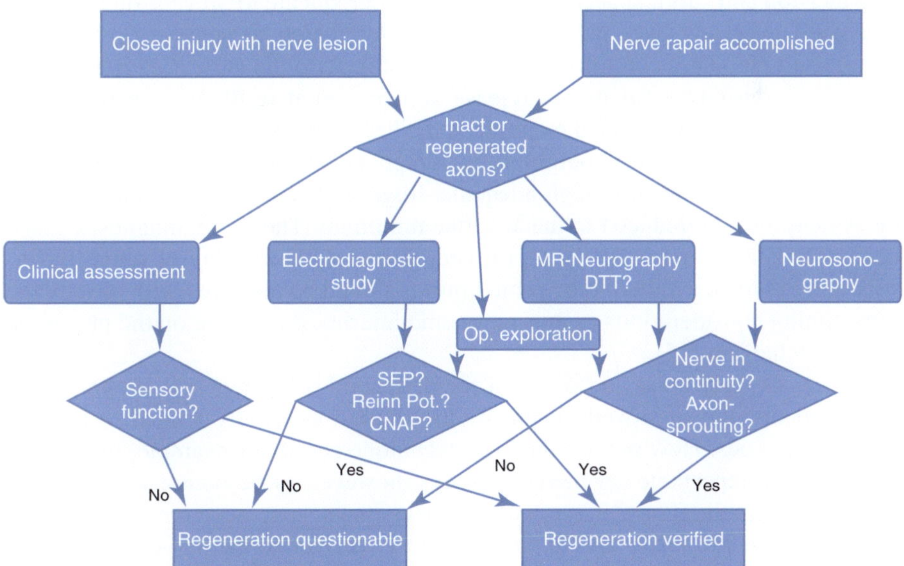

Fig. 3.5 Diagnostic procedures during follow-up

sprouting axons [20]. On the other hand, neurographic examinations and psychometric tests (e.g., Weinstein Enhanced Sensory Test, Semmes-Weinstein Monofilament Test, Shape-Texture Identification Test) are helpful for the documentation of the extent and degree of recovered sensitivity [18].

The techniques used within the revision surgery depend on the intraoperative findings and are basically not different from those used for the primary care (see Chap. 4). Whenever it is obvious to the surgeon that the nerve is irreversibly damaged without the chance of regeneration, he/she usually has to decide for one of the following options. In case the proximal nerve end is lacking, bypassing the defect using nerve fiber transfer from another nerve is an option. Direct muscle neurotization may be reasonable in case the distal stump is lacking. But also end-to-side coaptation or tendon transfers represent methods of choice. Especially in radial and peroneal nerve palsy, proper early tendon transfer – performed in parallel to the nerve reconstruction – could be considered because it is suitable to enhance wrist extension and grip function, i.e., foot lifting, while awaiting the return of nerve and muscle function.

3.9   Prevention of Nerve Injuries and Future Aspects

Although innovations in nerve reconstruction procedures have reached an excellent level, complete functional recovery is still rarely seen – except in children [16]. Therefore, it cannot be emphasized enough that it is of outmost importance to avoid injuries of peripheral nerves during any kind of surgical intervention [1]. Iatrogenic peripheral nerve injuries originate from disregardful approaches during skin incision and surgical access. An essential prerequisite to prevent such injuries is a good overview over the operation field for which to achieve; magnifying glasses and bloodless conditions could be helpful. Also the (so-called mini-) incisions, e.g., during carpal tunnel release surgery, often result in new peripheral nerve lesions. Another typical example is the lesion of the accessory nerve during disregardful extirpation of neighboring lymph nodes [7]. Whenever peripheral nerve lesions occur even though adequate diligence has been taken, the harm for the patient must be reduced to the absolute minimum. Therefore, diagnostic clarification has to be immediately performed, and prolonged delay of nerve reconstruction has to be avoided. Iatrogenic injuries should be managed with exactly the same timing considerations as non-iatrogenic injuries, depending on the presumed injury mechanism [15].

Taking the indication and decision for peripheral nerve repair is progressively facilitated through improved diagnostic techniques, especially related to modern imaging methods. How precise the DTT technique could, for example, predict the degree of an expectable regeneration keeps, however, to be demonstrated in the future. Once, in the future, it is possible to delay the degenerative processes related to peripheral nerve injury and to simultaneously accelerate the nerve regeneration processes, the timing of nerve repair will become less meaningful – the time period for successful surgical intervention would be significantly prolonged [11]. Furthermore, the decision processes related to nerve repair would be facilitated.

Despite the fact that current treatment strategies demonstrate some success, further efforts need to be done with the goal, to simultaneously potentiate axonal regeneration, increase neuronal survival, modulate central reorganization, and inhibit or reduce target organ atrophy [22].

References

1. Antoniadis G, Kretschmer T, Pedro MT, Konig RW, Heinen CP, Richter HP. Iatrogenic nerve injuries: prevalence, diagnosis and treatment. Dtsch Arztebl Int. 2014;111:273–9.
2. Assmus H. Somatosensory evoked potentials in peripheral nerve lesions. In: Barber C, editor. Evoked potentials. Lancaster: MTP Press limited; 1980. p. 437–42.
3. Assmus H. Sural nerve removal using a nerve stripper. Neurochirurgia (Stuttg). 1983;26: 51–2.
4. Birch R. Surgical disorders of the peripheral nerves. 2nd ed. New York: Springer; 2011. p. 231.
5. Brunelli G, Brunelli F. Strategy and timing of peripheral nerve surgery. Neurosurg Rev. 1990;13:95–102.
6. Brushart TM. Clinical nerve repair and grafting. In: Brushart TM, editor. Nerve repair. New York: Oxford University Press; 2011. p. 104–34.
7. Cesmebasi A, Smith J, Spinner RJ. Role of sonography in surgical decision making for iatrogenic spinal accessory nerve injuries: a paradigm shift. J Ultrasound Med. 2015;34:2305–12.
8. Chhabra A, Belzberg AJ, Rosson GD, Thawait GK, Chalian M, Farahani SJ, Shores JT, Deune G, Hashemi S, Thawait SK, Subhawong TK, Carrino JA. Impact of high resolution 3 tesla MR neurography (MRN) on diagnostic thinking and therapeutic patient management. Eur Radiol. 2016;26:1235–44.
9. Dahlin LB. The role of timing in nerve reconstruction. Int Rev Neurobiol. 2013;109:151–64.
10. Elton SG, Rizzo M. Management of radial nerve injury associated with humeral shaft fractures: an evidence-based approach. J Reconstr Microsurg. 2008;24:569–73.
11. Fowler JR, Lavasani M, Huard J, Goitz RJ. Biologic strategies to improve nerve regeneration after peripheral nerve repair. J Reconstr Microsurg. 2015;31:243–8.
12. Gordon T, English AW. Strategies to promote peripheral nerve regeneration: electrical stimulation and/or exercise. Eur J Neurosci. 2016;43:336–50.
13. Kretschmer T, Antoniadis G. Traumatische Nervenläsionen. In: Kretschmer T, Antoniadis G, Assmus H, editors. Nervenchirurgie. Heidelberg: Springer Berlin; 2014. p. 95–182.
14. Lundborg G. Alternatives to autologous nerve grafts. Handchir Mikrochir Plast Chir. 2004;36: 1–7.
15. Midha R. Mechanism and pathology of injury. In: Kim D, Midha R, Murovic JA, Spinner RJ, editors. Kline and Hudson's nerve injuries. 2nd ed. New York: Elsevier; 2008. p. 23–42.
16. Millesi H. Progress in peripheral nerve reconstruction. World J Surg. 1990;14:733–47.
17. Navarro X, Udina E. Chapter 6: methods and protocols in peripheral nerve regeneration experimental research: part III-electrophysiological evaluation. Int Rev Neurobiol. 2009;87: 105–26.
18. Rosen B, Lundborg G. Sensory re-education after nerve repair: aspects of timing. Handchir Mikrochir Plast Chir. 2004;36:8–12.
19. Simon NG, Spinner RJ, Kline DG, Kliot M. Advances in the neurological and neurosurgical management of peripheral nerve trauma. J Neurol Neurosurg Psychiatry. 2016;87:198–208.
20. Simon NG, Narvid J, Cage T, Banerjee S, Ralph JW, Engstrom JW, Kliot M, Chin C. Visualizing axon regeneration after peripheral nerve injury with magnetic resonance tractography. Neurology. 2014;83:1382–4.
21. Spinner RJ. Operative care and techniques. In: Kim D, Midha R, Murovic JA, Spinner RJ, editors. Kline and Hudson's nerve injuries. 2nd ed. New York: Elsevier; 2008. p. 87–105.
22. Tos P, Ronchi G, Geuna S, Battiston B. Future perspectives in nerve repair and regeneration. Int Rev Neurobiol. 2013;109:165–92.

Conventional Strategies for Nerve Repair

4

Mario G. Siqueira and Roberto S. Martins

Contents

> The central objective of nerve repair is to assist regenerating axons to re-establish useful functional connections with the periphery. Sir Sydney Sunderland

During the last decades, significant changes in the surgical management of nerve injuries have occurred, based on an improved knowledge of basic nerve biology and on the advance of surgical technologies like the use of magnification, bipolar coagulator, microinstruments, and fine suture material and the introduction of electrophysiologic methods for intraoperative assessment of nerve injuries. These advances led to improved functional results, increasing the number of surgical explorations and the attempts to repair lesions that previously were considered irreparable.

M.G. Siqueira, MD, PhD (✉) • R.S. Martins, MD, PhD
Peripheral Nerve Surgery Unit, Department of Neurosurgery,
University of São Paulo Medical School, Rua Virgilio de Carvalho Pinto,
381/apt.42, 05415-030 São Paulo, SP, Brazil
e-mail: mgsiqueira@uol.com.br

© Springer International Publishing AG 2017   41
K. Haastert-Talini et al. (eds.), *Modern Concepts of Peripheral Nerve Repair*,
DOI 10.1007/978-3-319-52319-4_4

In this chapter we describe the common surgical techniques in current use for nerve repair, external and internal neurolysis, end-to-end suture, and nerve grafting and two less used techniques, end-to-side suture and muscular neurotization.

4.1 Neurolysis

Neurolysis has been defined by Seddon [26] as an operation in which an injured nerve is freed from scar tissue or other neighboring tissue to facilitate regeneration. In this procedure whenever possible the tissue dissection should occur along anatomical planes. Attention should be devoted to hemostasis and minimal tissue damage, since bleeding and tissue debris will promote excessive scarring, which will attenuate the results of the surgical procedure.

There are two types of neurolysis, external and internal.

4.1.1 External Neurolysis

The external neurolysis consists of freeing the nerve from a constricting or distorting agent by dissection outside the epineurium, usually including the mesoneurium, an adventitious tissue that contains collateral blood vessels, and sometimes including the most external epineurium as well. The inner layers of the nerve remain intact. Nerve segments are freed circumferentially using a number 15 scalpel or Metzenbaum scissors. Seldom used as the treatment itself, the external neurolysis should be performed in all lesioned nerve segments before surgical reconstruction. It is usually begun by working from normal to abnormal nerve sections beginning well distal as well as proximal to the lesion site. Thickened or scarred portion of the external epineurium will then be resected. If carefully done, long lengths of nerves can be mobilized without serious interference with their blood supply. However, extensive manipulation may, in rare cases, promote neurological deterioration. A good deal of argument about the value of external neurolysis for the improvement of function in a direct fashion still exists. Apparently this technique could be valuable when the nerve is intact but tethered or immobilized by scar tissue and the patient complains of severe neuritic pain. In spite of this limited indication, external neurolysis is performed as the first step of almost all types of nerve repair. Figure 4.1 demonstrates the situation of a scarred sciatic nerve (Fig. 4.1a) and its appearance after external neurolysis (Fig. 4.1b).

4.1.2 Internal Neurolysis

Internal neurolysis is the exposure of nerve fascicles after epineurotomy and their separation by interfascicular dissection or by removal of interfascicular scar tissue. It is an essential part of some procedures [2] as follows: (1) separation of intact from damaged fascicles in partially damaged nerves, (2) separation of a fascicle during a

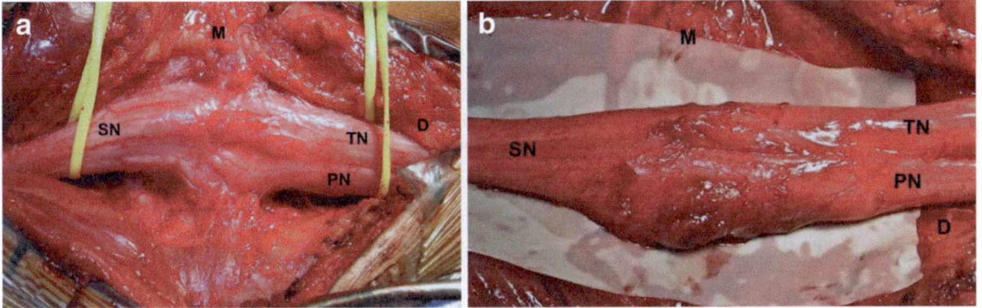

Fig. 4.1 Intraoperative view of a gunshot injury to the sciatic nerve. (**a**) Scar tissue involving the nerve. (**b**) After external neurolysis the main trunk of the nerve as well as its peroneal and tibial divisions can be identified. *D* distal, *M* medial, *PN* peroneal nerve, *SN* sciatic nerve, *TN* tibial nerve

nerve transfer, and (3) separation of intact fascicles during removal of a benign nerve tumor. Another indication for this procedure is given when there is an incomplete nerve tissue loss distal to the lesion, but the patient has pain of neuritic nature which does not respond to conservative management. When performing this technique, the surgeon should keep in mind that the removal of abundant fibrous tissue between the fascicles may impair the blood supply to this structure [18] with the potential risk of some loss of function.

4.2 End-to-End Neurorrhaphy

Since Hueter in 1873 [6] described an end-to-end coaptation of nerve ends by placing sutures in the epineurium, the end-to-end suture became the procedure of choice for nerve lesions where opposition of stumps can be gained without excessive tension. Excessive tension across a nerve repair site is known to impair the local blood circulation, to increase the scarring at the coaptation site, and finally to impair regeneration. The opposition of the nerve ends is facilitated by mobilization of the stumps, transposition of the nerve, and, in selected cases, mild flexion of the extremity. Every nerve repair should be performed with optical magnification (surgical loupes or microscope) and adequate lighting. The end-to-end neurorrhaphy should be done only when the nerve gap is small (usually less than 2 cm). A test to evaluate the possibility of direct suture without prohibitive tension involves passing an epineurial suture of 7-0 nylon. If the suture keep the stumps together without tearing the epineurium, the end-to-end neurorrhaphy is possible.

There are three types of end-to-end neurorrhaphy: epineurial, perineurial, and group fascicular. All three procedures always initiate with the preparation of the nerve ends. Transverse cuts, distant 1 mm from each other, are progressively made with a sharp instrument (micro scissors or surgical blade) until an area of healthy appearing fascicles without fibrotic tissue is reached. Following resection of the devitalized tissue hemostasis is imperative because bleeding could lead to excessive fibrosis and distortion of the nerve architecture. A small tipped bipolar coagulator or sponges dipped in a solution of 1:100,000 epinephrine in 10 ml of saline should be used for this purpose. The prepared nerve ends are then gently mobilized and approximated to be coapted, without excessive tension. In the *epineurial technique* the entire nerve trunk is sutured as a unit. Finely spaced interrupted nylon sutures inserted into the epineurium are used to approximate the stumps, first laterally and then along volar and dorsal epineurial surfaces. Suture material should be passed through the epineurium only, as the incorporation of neural elements results in scar tissue formation. The sutures should be placed approximately 0.5–1.0 mm from the incised edge, with the needle piercing the surface of the nerve and emerging just subepineurially. In the opposing nerve stump, the second passage of the needle begins subepineurially and emerges on the surface. The size, the depth, and the number of sutures should be minimized to decrease iatrogenic trauma and the formation of foreign body granuloma. The number of sutures required for adequate alignment of the stumps varies depending upon the nerve diameter. Having in mind that an adequate alignment is paramount for the success of the surgical procedure, it is desirable to perform the smallest number of sutures possible because all suture materials evoke an inflammatory reaction, which can result in production of excess granulation tissue. To maintain alignment of the nerve stumps, the first two sutures are placed in the nerve trunk 180° apart. Additional sutures are then placed in the upper portion of the nerve. Grasping carefully the ends of the two first sutures, the nerve trunk is rotated to expose the underside of the nerve where additional sutures are placed, completing the apposition of the nerve stumps. All sutures should be tied with equal tension. The tension applied should be just enough for alignment and contact of the neural bundles. Excessive tension may result in crushing and malalignment of the nerve bundles. Identification of the longitudinal epineurial blood vessels, which are not always present, helps to avoid rotation of the nerve ends and consequent malalignment of the fascicles. The visualization of fascicular patterns on the cut nerve surfaces can also be effective to help the correct realignment of peripheral nerve stumps in areas of consistent topography (e.g., distal ulnar and median nerves). The fascicular topography changes after 1–2 cm of neural trimming, but groups of fascicles can usually be opposed as closely as possible even though the repair is done at epineurial level. The epineurial technique is the most performed end-to-end neurorrhaphy in clinical practice and illustrated in Fig. 4.2.

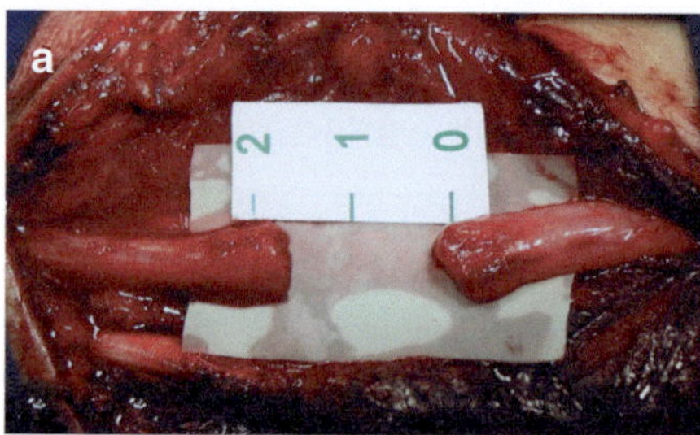
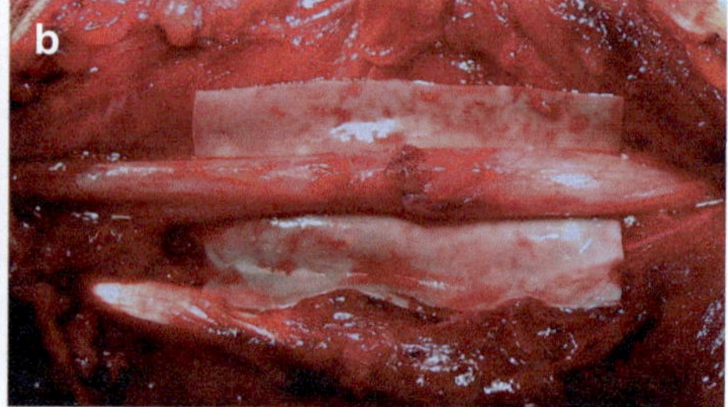

Fig. 4.2 Intraoperative view after resection of a neuroma in continuity of the ulnar nerve at the elbow. (**a**) Distance between the two stumps of the nerve after resection of the lesioned tissue and normal retraction (*nerve gap*). (**b**) End-to-end epineurial repair

As expertise and technical development in microsurgery have progressed, suture repair of peripheral nerve subunits, like the *perineurial or fascicular repair*, has increased in popularity. The technique involves resection of the outer epineurium, followed by intraneural dissection of fascicles in both nerve stumps and perineurial suturing of individual fascicles with one or two sutures of 10-0 suture material. The perineurial repair represents the best possibility of nerve alignment by the surgeon. However, this advantage may be offset by the amount of neural trauma the technique demands. In clinical practice this procedure is seldom performed. *Grouped fascicular repair* is a less aggressive method of nerve alignment done by the identification of grouped fascicular patterns in both nerve ends and suturing through the thickened inner epineurium. This technique is used mainly in areas of well-defined nerve topography such as the distal median and ulnar nerves and the radial nerve around the elbow.

Superiority of one end-to-end neurorrhaphy over another has never been clearly demonstrated [8]. In practice, the accurate alignment of the fascicles or grouping of fascicles is often difficult because of trauma, edema, and scarring that can distort the normal topography.

4.3   End-to-Side Neurorrhaphy

End-to-side neurorrhaphy was first described by Letievant in 1873 [30], but the idea was abandoned due to poor results. More than a century later, Viterbo et al. [32] reintroduced the technique with apparently promising results. The end-to-side neurorrhaphy involves coaptation of the distal stump of a transected nerve to the trunk of an adjacent healthy donor nerve. It has been proposed as an alternative technique when the proximal stump of an injured nerve is unavailable or when the nerve gap is too long to be bridged by a nerve graft. Collateral sprouting is the accepted mechanism of nerve regeneration following end-to-side neurorrhaphy, where regenerating axons originated from the most proximal Ranvier's node of the donor nerve grow toward the coaptation site [29, 37]. Whether the receptor nerve should be coapted to the donor nerve through an epineurotomy or a perineurotomy is still controversial. Although some experimental papers revealed no difference if a nerve window at the coaptation site was made or not [32, 33], other investigators claim that the greater degree of axonal damage to the donor nerve after a perineurotomy enhances axonal regeneration with better histological results [35, 36]. The clinical experience with this technique has been published in the form of case reports and small clinical series, and no randomized clinical trials have been performed in order to compare end-to-side coaptation to other reconstructive techniques. The clinical outcomes of end-to-side repair are often disappointing. In a recent published review of the clinical applications of the technique, Tos et al. [30] demonstrated that a discrepancy between experimental and clinical results still exists, and the authors concluded that at present the end-to-side repair could not substitute standard techniques in most situations. In the majority of cases, it will provide only limited sensory recovery [23, 28, 34]. It can be considered a valid therapeutic option only in cases of failure of other attempts of nerve repair or whenever other approaches are not feasible, especially when protective sensibility is a reasonable goal.

4.4 Graft Repair

Nerve grafting dates back to Philippeaux and Vulpian in 1817 [9]. In extensive injuries, especially those due to blunt mechanisms, loss of nerve tissue may produce lengthy lesions which, when resected, result in a large nerve gap. A nerve gap is defined as the distance between two ends of a severed nerve and consists not only of an amount of nerve tissue lost in the injury or debridement but also of the distance that the nerve has retracted due to its elastic properties [15]. Small nerve gaps (<2 cm) can be overcome by stretching the nerve stumps to a limited extent to attain apposition, making possible a primary repair. But when a significant amount of stretching and mobilization is necessary, the consequent increase in the suture line tension endangers the extrinsic vascular supply to the nerve leading to connective tissue proliferation and formation of scar tissue [13]. In this situation of an irreducible nerve gap, the gold standard management continues to be autologous nerve grafting. The nerve grafts serve as a guide for the axons of the proximal stump as it regrows toward the distal stump.

Small-caliber grafts seem to serve better than longer whole nerve grafts [14]. For a nerve graft to survive, it must be revascularized, and when the nerve is too thick, the central part of the nerve graft will not become revascularized, and the outcome of the repair will be poor. The sural nerve, by far the most commonly used donor nerve, is harvested from the ankle until near the knee, and 30–40 cm of the nerve is usually obtained in adults from each leg for grafting. Other sensory nerves like the medial antebrachial cutaneous or the sensory branch of the radial nerve are used as well. The grafts should be harvested after the injured nerves have been exposed, the extent of lesion defined, and the gap between the prepared nerve stumps measured. Then the number of grafts required is calculated. To release tension on suture lines, the length of the grafts should be about 15–20% greater than the measured gap because they always present some shrinkage owing to a relative initial hypovascularization. The nerve grafts are initially similar to other devascularized tissue implants. The regeneration of the blood supply must be provided by the nerve stumps and surrounding tissues and takes some time. In the beginning the graft relies on the imbibition from the surrounding for nutrition. Consequently, long grafts and a poorly vascularized tissue bed could be responsible for ischemic necrosis of the central graft core, with destruction of Schwann cell tubules and failure of axonal regeneration through the graft.

The most popular as a graft technique is an interfascicular grouped fascicular approach described by Millesi in the early 1970s [16, 17]. The principles and surgical technique of nerve grafting are similar to direct repair. The proximal and distal ends of the nerve are transversely sectioned until viable fascicles are visualized, and groups of fascicles are then isolated both proximally and distally. Usually oriented in a reverse fashion to minimize the diversion of regenerating fibers from the distal neurorrhaphy, a number of small-caliber nerve grafts are attached between the nerve ends, connecting corresponding groups of fascicles. The coaptation is maintained by one or two fine sutures often supplemented by fibrin glue. As much of the

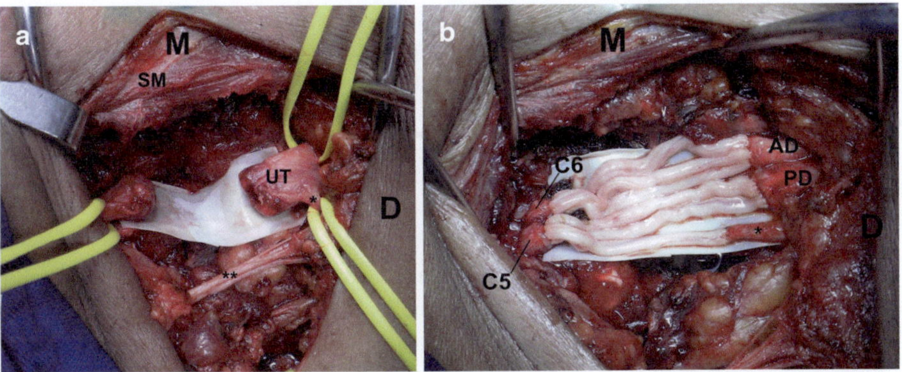

Fig. 4.3 Intraoperative view of a penetrating stab wound to the right supraclavicular region. (**a**) An injury of the upper trunk of the brachial plexus was identified. (**b**) Reconstruction was performed with nerve grafts. *AD* anterior division of the upper trunk, *C5* fifth spinal nerve, *C6* sixth spinal nerve, *D* distal, *M* medial, *PD* posterior division of the upper trunk, *SM* sternocleidomastoid muscle, *UT* upper trunk, * suprascapular nerve, ** supraclavicular nerves

fascicular structure of each stump as possible is covered in this fashion. Individual grafts should be positioned loosely, not too close to each other, to permit maximal contact with a viable recipient bed. Figure 4.3 illustrates a brachial plexus injury that has been repaired by an interfascicular grouped fascicular approach.

Graft length might influence regeneration as longer grafts may be harder to revascularize, but in clinical practice no agreement exists on the maximal length that may be bridged by a nerve graft. Although good results are eventually reported with longer grafts, most nerve surgeons agree that the outcome is worse with grafts greater than 10 cm.

4.5 Direct Muscular Neurotization

Described in the beginning of the twentieth century, the surgical insertion of peripheral nerves directly into denervated muscles is called direct muscular neurotization. This procedure is indicated when no distal nerve stump is available for neural coaptation or when the lesion involves the neuromuscular junction [24]. Experimentally it was observed that the implantation of a normal nerve near denervated motor end plates reinnervates this site and that axons that do not have contact with those persistent motor end plates will induce new ones in previously denervated areas [22]. However, clinically the neurotization restores significantly less function, when compared with direct repair or grafting, leaving areas of the target muscle denervated [12]. In most published reports, an entire nerve was implanted into the target muscle, probably leaving denervated areas outside the reach of the regenerating axons. To overcome this problem, Brunelli [4] suggested that the donor nerve should be splitted into multiple fascicles and distributed widely across the muscle. Direct muscular neurotization is a potentially effective technique when the normal nerve-muscle interface has been destroyed [1], but until now there are only a few reports of clinically successful reinnervation in the literature, and this technique has no stablished role in reconstructive nerve surgery.

4.6   Fibrin Glue Versus Suture

Epineurial suture repair is generally considered as the gold standard for peripheral nerve repair, but when nerve trauma is extensive, the suture method can be difficult and time-consuming. Specific training is necessary for nerve repair by suture which requires the placement of stitches that persist as foreign bodies producing inflammation and different degrees of scarring. The number of stitches (the less the better [11]) and the surgical skill certainly play a role in the improvement of outcomes.

Fibrin glue is one of the alternatives to suture [10]. Concentrated fibrinogen and thrombin are common ingredients in the mostly used fibrin glues, which are differing in the antifibrinolytic agent contained or the application procedure. Currently, the use of fibrin sealants as nerve glue still has not been approved and their use on nerve surgery is considered off-label.

The fibrin sealants simulate the last stages of the clotting cascade forming a substance resembling a physiologic blood clot that holds the nerve ends together [7]. The artificial "clot" protects the repair from scar tissue and allows healing to occur. Its structural integrity is preserved for about 3 weeks by the antifibrinolytic component of the sealant [5].

The potential advantages of fibrin glue for nerve repair include ease of use, reduced operative time, less tissue manipulation/trauma with consequent less inflammation/fibrosis, and maintenance of nerve architecture with better fascicular alignment [3, 19, 20, 27].

The amount of publications concerning the use of fibrin sealants as nerve glue is small. A recently published systematic review [25] found 14 animal studies, one cadaver study, and only one clinical study that fit the study criteria. Although some of the results were conflicting, most found fibrin glue repair to be efficient (and sometimes even superior) to suture repair.

The following are some practical remarks: (1) Nerve repair with fibrin glue has an initial low tensile strength, and its use should be limited to situations without tension in the coaptation (grafts and nerve transfers) and in cases with difficult exposures or exceptionally small-caliber nerves; (2) Before the use of the glue, a meticulous hemostasis should be done, and the nerve surfaces should be dry of excess fluids to ensure optimal adherence [31]; (3) After nerve repair, the nerve glue should be left to polymerize and cross-link for several minutes before irrigation; (4)The inevitable small amount of glue that stays between the nerve ends should not be a concern as fibrin glue is nontoxic and does not block axon regeneration [21]; (5) Like in the repair by suture, at the end of the surgery, the upper extremity should be immobilized for 3 weeks to ensure an ideal environment for axon regeneration.

Despite the apparent advantages of fibrin glue, its low tensile strength should always be kept in mind. To overcome this potential disadvantage, two combined strategies were created: to add fibrin glue to a standard suture repair and to reduce the number of stitches by using fibrin glue to reinforce the repair. There is no advantage with the first strategy, but the reduced number of stitches may ultimately lead to better outcomes. In practice the use of a few stitches complemented by the fibrin glue to enhance the coaptation has been adopted by many nerve surgeons.

4.7   Factors Influencing the Results of Surgical Repair

Besides the surgical techniques, the results of repair of peripheral nerves are certainly influenced by some biological aspects:

1. Younger patients recover more completely and in a shorter period of time. This is probably related to shorter limb length, faster rate of regeneration, and greater adaptability and compensatory sensory and motor reeducation.
2. The level of the injury. Too proximal injuries require greater metabolic biosynthesis for functional return, and this may exceed the capabilities of the nerve cell body and result in cell death.
3. The result of the repair of pure motor or pure sensory nerves is usually better. In mixed-function nerves, the potential for transposition of axons during regeneration with improper end-organ reinnervation exists.
4. The extent of the injury. Lesions in continuity or those with focal neuroma formation or small gaps will present better results than injuries with irreducible gaps or long length of defect owing to segmental vascular supply, suture line tension, and biologic considerations in nerve grafting.
5. Associated injury may add further difficulty to nerve regeneration. Polysystemic trauma, massive deep wounds, sepsis, scar formation, and contraction wound healing may interfere with the management of the patient.
6. The merits and indications for immediate versus early secondary repair have been discussed in Chap. 3. However, it is important to emphasize that as the interval between the time of injury and surgical intervention increases, irreversible changes occur in the nerve trunk, particularly in the distal segment. In addition, neurogenic atrophy and fibrosis of denervated muscle segments complicate the potential for functional recovery. Therefore, early repair is advocated, whenever possible.

References

1. Becker M, Lassner F, Fansa H, Mawrin C, Pallua N. Refinements in nerve to muscle neurotization. Muscle Nerve. 2002;26:362–6.
2. Birch R. Surgical disorders of the peripheral nerves. 2nd ed. London: Springer; 2011.
3. Boedts D. A comparative experimental study on nerve repair. Arch Otorhinolaryngol. 1987;244:1–6.

4. Brunelli G. Direct neurotization of severely damaged muscles. J Hand Surg [Am]. 1982;7: 572–9.
5. Dunn CJ, Goa KL. Fibrin sealant: a review of its use in surgery and endoscopy. Drugs. 1999;58:863–86.
6. Ijkema-Paassen J, Jansen K, Gramsbergen A. Transection of peripheral nerves, bridging strategies and effect evaluation. Biomaterials. 2004;25:1583–92.
7. Isaacs JE, McDaniel CO, Owen JR, Wayne JS. Comparative analysis of biomechanical performance of available "nerve glues". J Hand Surg [Am]. 2008;33:893–9.
8. Lundborg GA. A 25-year perspective of peripheral nerve surgery: evolving neuroscientific concepts and clinical significance. J Hand Surg [Am]. 2000;25:391–414.
9. Mackinnon S, Dellon AL. Surgery of the peripheral nerve. New York: Thieme Medical; 1988.
10. Martins RS, Siqueira MG, Da Silva CF, Plese JP. Overall assessment of regeneration in peripheral nerve lesion repair using fibrin glue, suture, or a combination of the 2 techniques in a rat model: which is the ideal choice? Surg Neurol. 2005;64:10–6.
11. Martins RS, Teodoro WR, Simplício H, Capellozi VL, Siqueira MG, Yoshinari NH, Pereira JP, Teixeira MJ. Influence of suture on peripheral nerve regeneration and collagen production at the site of neurorrhaphy: an experimental study. Neurosurgery. 2011;68:765–72.
12. McNamara MJ, Garrett WE, Seaber AV, Goldner JL. Neurorrhaphy, nerve grafting, and neurotization: a functional comparison of nerve reconstructive techniques. J Hand Surg [Am]. 1987;12:354–60.
13. Millesi H. Healing of nerves. Clin Plast Surg. 1977;4:459–73.
14. Millesi H. Reappraisal of nerve repair. Surg Clin North Am. 1981;61:321–40.
15. Millesi H. The nerve gap: theory and clinical practice. Hand Clin. 1986;2:651–63.
16. Millesi H, Meissl G, Berger A. The interfascicular nerve-grafting of the median and ulnar nerves. J Bone Joint Surg Am. 1972;54:727–50.
17. Millesi H, Meissl G, Berger A. Further experience with interfascicular grafting of the median, ulnar, and radial nerves. J Bone Joint Surg Am. 1976;58:209–18.
18. Millesi H, Rath TH, Reihsner R, Zoch G. Microsurgical neurolysis: its anatomical and physiological basis and its classification. Microsurgery. 1993;14:430–9.
19. Narakas A. The use of fibrin glue in repair of peripheral nerves. Orthop Clin North Am. 1988;19:187–99.
20. Ornelas I, Padilla L, Di Silvio M, Schalch P, Esperante S, Infante PL, Bustamante JC, Avalos P, Varela D, López M. Fibrin glue: an alternative technique for nerve coaptation – part I. Wave amplitude, conduction velocity, and plantar-length factors. J Reconstr Microsurg. 2006;22: 119–22.
21. Palazzi S, Vila-Torres J, Lorenzo JC. Fibrin glue is a sealant and not a nerve barrier. J Reconstr Microsurg. 1995;11:135–9.
22. Payne SH, Brushart TM. Neurotization of the rat soleus muscle: a quantitative analysis of reinnervation. J Hand Surg [Am]. 1997;22:640–3.
23. Pienaar C, Swan MC, De Jager W, Solomons M. Clinical experience with end-to-side nerve transfer. J Hand Surg Br. 2004;5:438–43.
24. Sakellarides H, Sorbie C, James L. Reinnervation of denervated muscles by nerve transplantation. Clin Orthop. 1972;83:194–201.
25. Sameen M, Wood TJ, Bain JR. A systematic review on the use of fibrin glue for peripheral nerve repair. Plast Reconstr Surg. 2011;127:2381–90.
26. Seddon HJ. Surgical disorders of the peripheral nerve. 2nd ed. Churchill-Livingstone: Edinburgh; 1975.
27. Smahel J, Meyer VE, Bachem U. Glueing of peripheral nerves with fibrin: experimental studies. J Reconstr Microsurg. 1987;3:211–20.
28. Tarasidis G, Watanabe O, Mackinnon SE, Strasberg SR, Haughey BH, Hunter DA. End-to-side neurorrhaphy: a long-term study of neural regeneration in a rat model. Otolaryngol Head Neck Surg. 1998;119:337–41.
29. Tham SK, Morrison WA. Motor collateral sprouting through an end-to-side nerve repair. J Hand Surg [Am]. 1998;23:844–51.

30. Tos P, Colzani G, Ciclamini D, Titolo P, Pugliese P, Artiaco S. Clinical applications of end-to-side neurorrhaphy: an update. Biomed Res Int. 2014;2014:646128.
31. Tse R, Ko JH. Nerve glue for upper extremity reconstruction. Hand Clin. 2012;28:529–40.
32. Viterbo F, Trindade JC, Hoshino K, Mazzoni NA. Lateroterminal neurorrhaphy without removal of the epineural sheath. Experimental study in rats. Rev Paul Med. 1992;110: 267–75.
33. Viterbo F, Trindade JC, Hoshino K, Padovani CR. End-to-side neurorrhaphy with and without perineurium. São Paulo Med J. 1998;116:1808–14.
34. Yuksel F, Peker F, Celikoz B. Two applications of end-to-side nerve neurorrhaphy in severe upper-extremity nerve injuries. Microsurgery. 2004;24:363–8.
35. Zhang Z, Soucacos PN, Beris AE, Bo J, Joachim E, Johnson EO. Long term evaluation of rat peripheral nerve repair with end-to-side neurorrhaphy. J Reconstr Microsurg. 2000;16: 303–11.
36. Zhang Z, Soucacos PN, Bo J, Beris AE, Malizos KN, Joachim E, Agnantis NJ. Reinnervation after end-to-side nerve coaptation in a rat model. Am J Orthop. 2001;30:400–6.
37. Zhao JZ, Chen ZW, Chen TY. Nerve regeneration after terminolateral neurorrhaphy: experimental study in rats. J Reconstr Microsurg. 1997;13:31–7.

Surgical Techniques in the Lesions of Peripheral Nerves

5

Kartik G. Krishnan

Contents

5.1 Introduction

Some basic prerequisites for the successful performance of any surgical operation are the following: (a) profound knowledge [of the structures being manipulated—anatomy, physiology, biochemistry, pathophysiology, eventual anomalies, and the course of abnormalities with passage of time], (b) adequate exposure of the structures being manipulated [clear illumination, optical magnification, and the variety of surgical approaches to the target structures], (c) familiarity and dexterity with the tools of surgery and intraoperative diagnostics, and (d) a sound understanding of the goals to be achieved, the means to achieve them, the merit the procedure might bring forth, and implementation of this knowledge to the situation at hand. The aforementioned concepts do not stand alone as separate entities; they rather intertwine in the most appropriate fashion as the surgeon delivers his service to achieve a particular goal he has in mind.

K.G. Krishnan, MD, PhD
Division of Reconstructive Neurosurgery, Justus Liebig University,
Klinikstrasse 33, D-35929 Giessen, Germany
e-mail: Kartik.Krishnan@neuro.med.uni-giessen.de

© Springer International Publishing AG 2017
K. Haastert-Talini et al. (eds.), *Modern Concepts of Peripheral Nerve Repair*,
DOI 10.1007/978-3-319-52319-4_5

In the given context, let us consider the example of a peripheral nerve lesion at its early stage requiring a surgical procedure. One is confronted with specific tasks of (a) understanding the intrinsic nature of the lesion itself and the functional derangement it has brought forth; (b) decision-making about the choice of the method of surgical treatment, viz., release the nerve or reconstruct it with grafts or other means; (c) deciding the choice of technique and surgical approach to the lesion; and (d) implementing intraoperative diagnostics to deepen the understanding of the lesion.

This chapter will narrate the technical aspects of surgery for peripheral nerve lesions.

5.2 Technology of Illumination and Optics

Like in any other discipline, the surgical approach to peripheral nerve lesions requires adequate exposure of the anatomical structures, excellent illumination of the operating field, and magnification [14, 15]. There are various ways to achieve this.

5.2.1 The Operating Microscope

The modern operating microscope had a widely rotatable head equipped with a cold light and an optical beam splitter. The microscope head is held by a delicately balanced holding system, which is either stationary (ceiling mounted) or mobile (wheel mounted). The modern operating microscope (Fig. 5.1) comes with

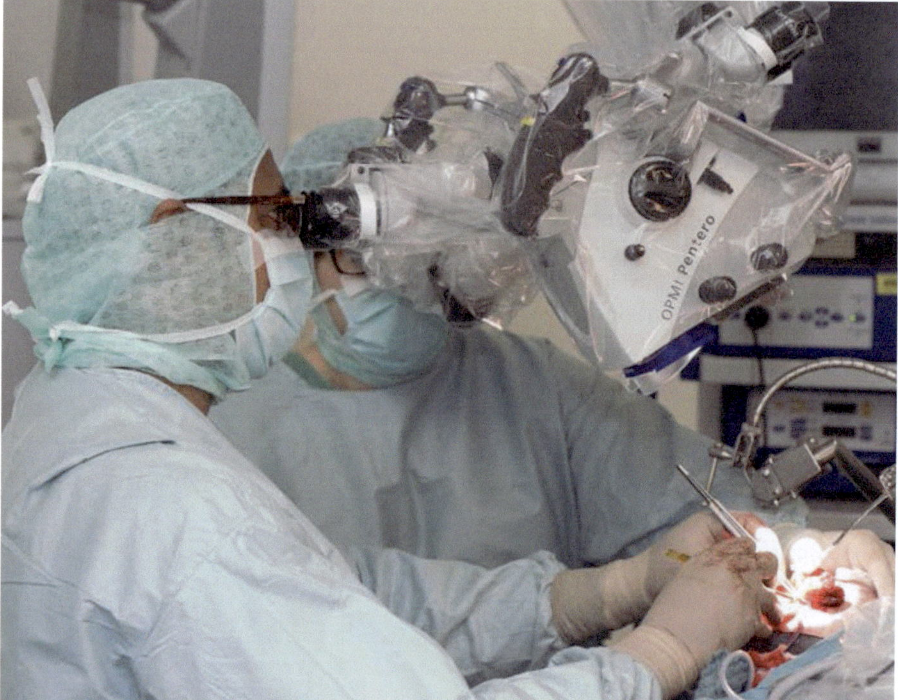

Fig. 5.1 The modern operating microscope with the optical beam splitter

integrated changeable optic filters for fluorescence microscopy and navigation match possibilities that aid in additional intraoperative diagnostics and tracking, respectively. In modern times, the operating microscope is an integral part of any surgical procedure. Microminiature sutures of nerves and vessels are best done under microscopic magnification (as opposed to magnification using binocular loupes). Perfusion of blood vessels and patency of anastomoses are best studied using indocyanine green video angiography integrated to the operating microscope. The one major disadvantage of the operating microscope today is its physical bulk. It simply takes away much valuable space in the operating setting.

5.2.2　The Retractor-Integrated Endoscope

Minimally invasive surgical methods are becoming increasingly popular. Post-traumatic nerve lesions are frequently, if not always, associated with severe scarring; it is advisable to widely expose such scarred post-traumatic nerve lesions. However, compression neuropathies and other pathological changes not accompanied by scarring may and can be explored through minimal skin incisions using the endoscope [9, 12, 13]. Formerly endoscopic surgery was limited to predefined body spaces such as paranasal sinuses, thoracic cavity, abdominal cavity, joint spaces, etc. The one exception of the use of endoscopy in nerve surgery was the carpal tunnel syndrome, where the tight canal was blindly dilated using blunt bougies in order to introduce the endoscope. This method is available both as monoportal and as biportal techniques, however limited to the decompression of the carpal tunnel [5]. Beginning of this century, we designed the retractor-integrated endoscope named after the author, in cooperation with the Karl Storz Company of Tuttlingen, Germany (Fig. 5.2), which is a universal tool for use on any nerve on the body [12, 13].

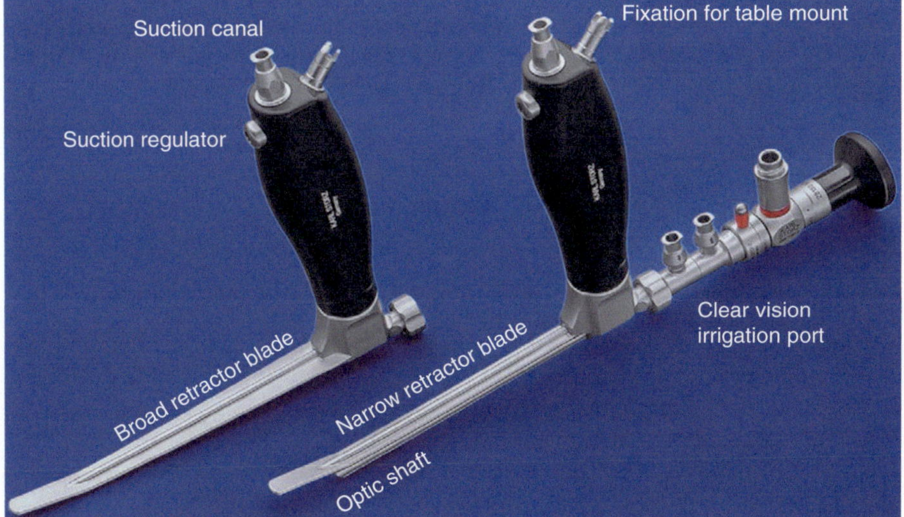

Fig. 5.2 The retractor-integrated Krishnan endoscope in two variations. The narrow one is used for releasing the median nerve at the carpal tunnel and the tibial nerve at the tarsal tunnel. For all other purposes, the broader blade is used

The principle is to create a space along the topographical course of the nerve [or any other structure of surgical interest] by means of soft-tissue retraction and manipulate the nerve [or the structure of interest]. This technique has found a wide range of indications and routine application in simple decompression and transposition of peripheral nerves irrespective of their anatomical location, extensive exploration of nerves for occult pathology, simple nerve suturing, and even harvesting nerves for grafting. The one main disadvantage of the retractor-integrated endoscope is its limitation for use only in non-scarred regions. Furthermore, the use of the retractor endoscope for the exploration of nerves in patients with a rich layer of subcutaneous adipose tissue requires extraordinary skills. Clinical trials have shown the feasibility of application of the retractor endoscope for the exploration of almost any nerve of the extremities [9, 12]. Figure 5.3 exemplarily depicts the decompression of the median nerve in carpal tunnel syndrome using the retractor endoscope. Trials comparing the open nerve release with the endoscopic release have shown that the long-term results of both methods are just the same; however, the short-term results of the retractor endoscopic nerve decompression are superior to the open technique [4].

One issue of endoscopic exploration of peripheral nerves worthy of mention here is the use of tourniquets on extremities. Application of exsanguinating or non-exsanguinating tourniquets at the proximal part of the extremity highly facilitates recognition and visualization of anatomical structures. However, improper application of very high pressures for prolonged periods might result in secondary iatrogenic compression neuropathies and might prove counterproductive. It is to be borne in mind that there is no single empirical pressure level for upper and lower extremities. My preferred method is to add 80–110 mm Hg to the present systolic pressure and pump up the tourniquet to that value. For example, I will apply 180 mm Hg tourniquet pressure (in a person with a thin arm) when the present systolic pressure is 100 mm Hg. In a person with abundant subcutaneous fat tissue with a systolic pressure of 100 mm Hg, I will recommend a tourniquet pressure of no more than 220 mm Hg. Blindly pumping up to 300 mm Hg for arms and 400 mm Hg for legs should be strongly discouraged. Tourniquets nullify the possibility of any and all electrophysiological measurements. Thus tourniquets should not be used, when one contemplates intraoperative monitoring or diagnostics.

5.2.3 The Video Telescope Operating Microscopy (ViTOM) or the Exoscope

The ViTOM telescope is an exoscope that was designed as an additional observation tool complementing the microscope in open surgical procedures [17, 20]. Being a derivative of the endoscope, the exoscope optic is held in place at a distance of 25–75 cm from the object of interest using a simple mechanical holding arm, the endoscope camera and the light cable are connected to the exoscope at the provided

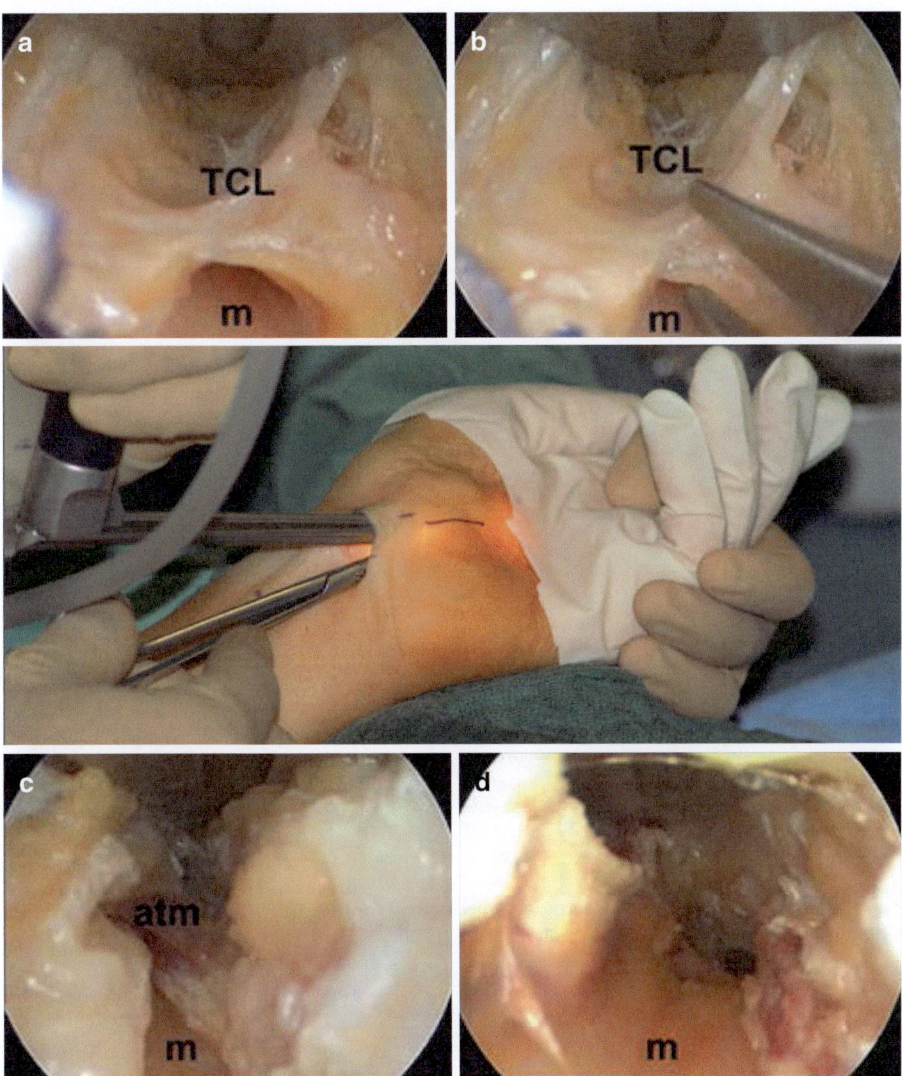

Fig. 5.3 The retractor endoscopic decompression of the carpal tunnel. (**a**–**d**) The steps of the surgery until the transverse carpal ligament (*tcl*) is transected, and the median nerve (*m*) is deroofed along its course within the carpal tunnel, *atm* accessory thenar muscle

slots, and the high-definition image is projected to an external monitor (Fig. 5.4). The camera offers an optical magnification of 1–2×. The effective magnification achieved with the exoscope depends on the working distance and the size and resolution of the monitor used. For example, with a minimal working distance of 25 cm, the object field of approximately 3.5 cm is achieved with a 2× zoom of the camera;

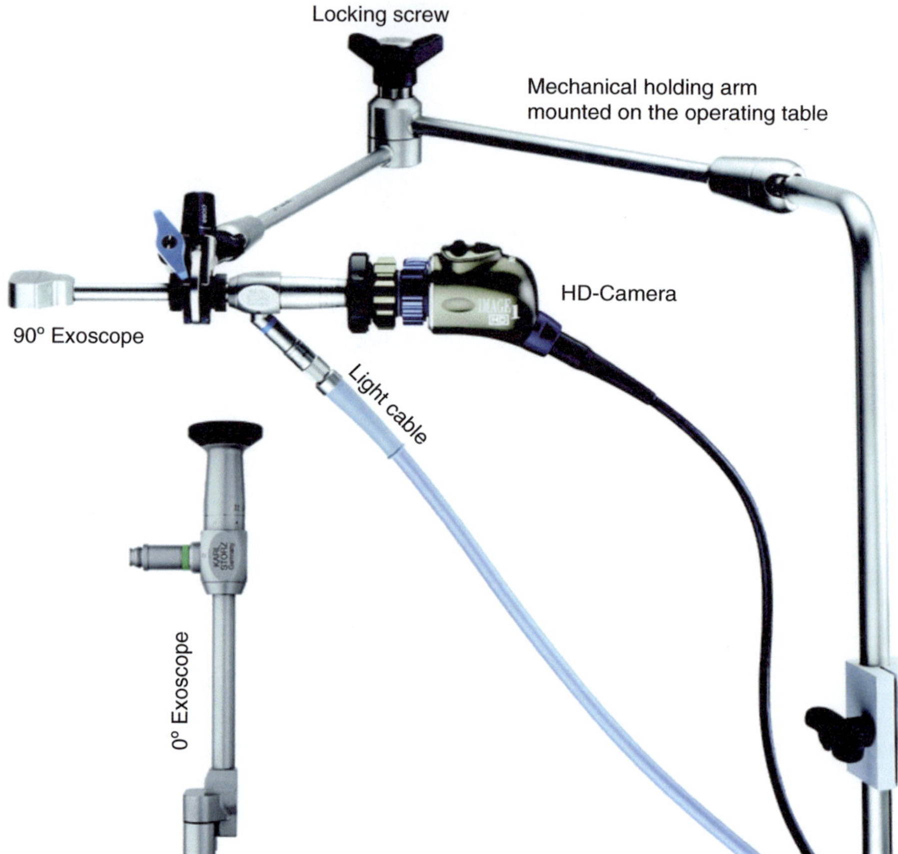

Fig. 5.4 The setting of the exoscope. The mechanical arm is shown to hold the 90° exoscope, whereas the 0° exoscope is shown as an inset

a 26″ monitor will offer a maximal effective magnification of 16×, whereas a 52″ monitor is capable of offering a 34× magnification. Encouraged by the sleekness of the system, several groups, including ours, studied the application of the exoscope, where usually one would use an operating microscope. In this feasibility study, we successfully performed lumbar spinal discectomies, anterior cervical discectomies and fusion, evacuation of intracerebral hematomas, removal of schwannomas from peripheral nerves, and even microvascular anastomoses and microneural sutures [8]. A possible surgical setup for exoscopic surgery is shown in Fig. 5.5.

The major disadvantage of the exoscope is its mechanical holding arm and the cumbersome refocusing and variation of magnification. This disadvantage has more to do with the holding system, rather than the optics and illumination offered by the exoscope. Various alternative holding arm systems are available, Endocrane, UniArm, Point Setter, and Versacrane, to name a few. Important features that are yet to be integrated with the exoscope are fluorescence microscopy and navigation

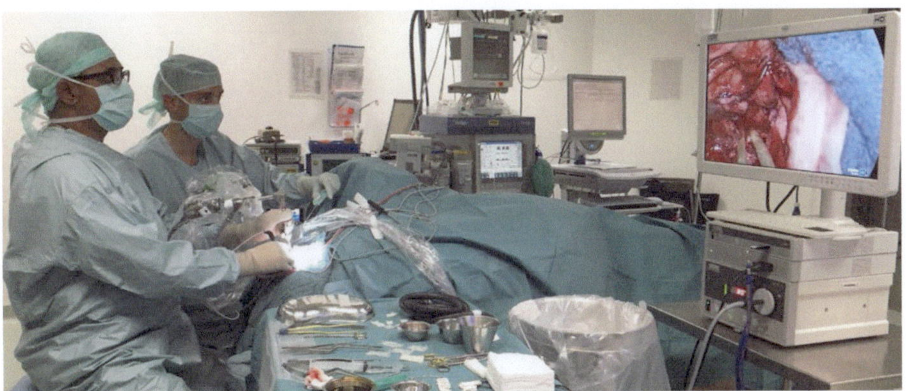

Fig. 5.5 A surgical procedure performed under HD-exoscopic illumination and magnification

match, whereas endoscopes already offer these prospects. In the recent years, augmented reality and image superimposition technology have shown rapid evolution and are put to use in the automobile industry. The magnifying high-definition exoscope, especially when integrated with such powerful tools, is capable of evolving into yet another advanced gadget for performing surgical operations.

Irrespective of the technology used for achieving illumination, magnification, and exposure, these appliances should be seen as tools in the armamentarium of the contemporary surgeon. Routine use of such technology will make the surgeon aptly recognize the indications for their application whenever and wherever found appropriate.

5.3 Techniques in Nerve Coaptation

Allegedly, the first reported nerve coaptation was performed by the celebrated Persian physician Avicenna. Before his times peripheral nerves belonged to the category of "noli me tangere" or "touch me not," due to a false conception that touching severed nerves produced epileptic seizures. Avicenna himself advocated not touching the nerve, rather bring its severed ends together by adapting the surrounding connective tissue [15]. The results of axonal growth depend directly on the amount of foreign [suture] material implanted to perform the nerve coaptation. Consequently, meticulous microsurgical techniques were employed, and neurorrhaphies came to be performed using microminiature suture material (Fig. 5.6). In order to achieve precision in coaptation of the individual proximal fascicles to their distal counterparts, the interfascicular nerve suturing technique became popular. However, this was quickly discarded, owing to the amount of tissue scarring the interfascicular suture technique had caused within the repaired nerve. The contemporary nerve repair technique involves the epineural adaptation of the nerve as a whole using a few microsutures, having provided the correct orientation of the proximal and distal stumps, and buttressing the suture with the application of fibrin glue

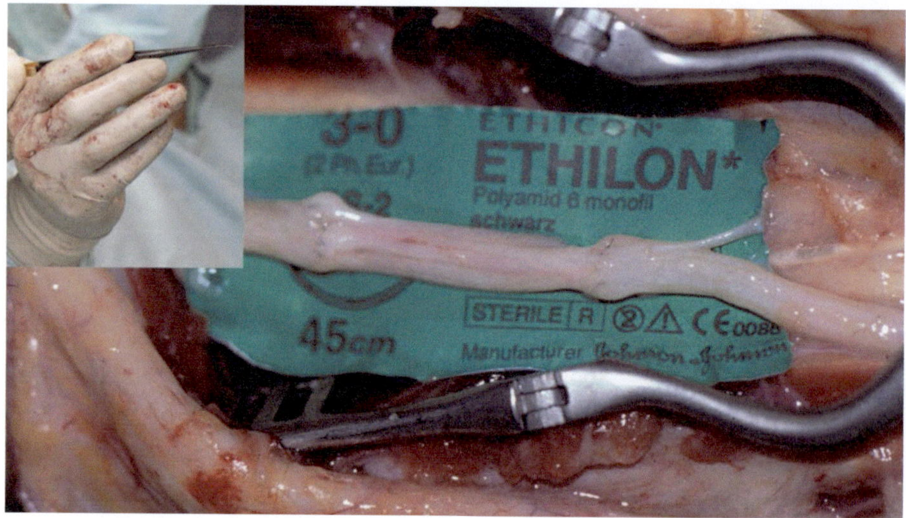

Fig. 5.6 Tension-free and torque-free microminiature suture of the median nerve with grafts. The inset shows the surgeon's hand holding a needle driver bearing the suture material

[which is absorbed quickly and replaced by a fine film of autologous fibrous sheath] (also see Chap. 4). Some experimental works have tested suture-free methods such as laser welding of nerve ends, which have somehow failed to enter the main stream of clinical practice [2, 18]. A detailed description of microsurgical techniques and placement of microminiature suture of nerves are out of the scope of this chapter, and the reader is referred to Krishnan [10].

One other important issue to prevent [or minimize] intraneural scarring is to attain a tension-free coaptation of the nerve ends. Consequently, the higher the tension at the suture line, the poorer is the axonal sprouting and growth across that suture line. Thus, it is agreed that injury of nerves with tissue deficiency are better grafted than sutured under tension. Peripheral nerve grafts are taken from superficial sensory nerves, some of them being sural nerves, saphenous nerves, and medial and lateral antebrachial cutaneous nerves. It is to be borne in mind that while grafting a "nerve," the surgeon does not graft axons, rather Schwann cells that serve as a scaffolding for the sprouting and growing axons from the proximal nerve stump in order to reach its target structure located distally. There has been a profound and lasting search for equipotential alternatives to autologous nerve grafts—however so far in vain. Chapter 7 in this book deals with this issue. Further developments are awaited along these lines (Chap. 10).

The technical aspects of harvesting nerves for grafting deserve some mention in this chapter dedicated to surgical technique. The classical approach to harvesting a nerve is to make the skin incision along the entire course of the graft, carefully dissect and transect the nerve, and prepare it for autologous transplantation. However, this approach makes the nerve harvest a cumbersome major operation in its own right. In order to minimize the trauma of nerve harvest, endoscopic techniques were

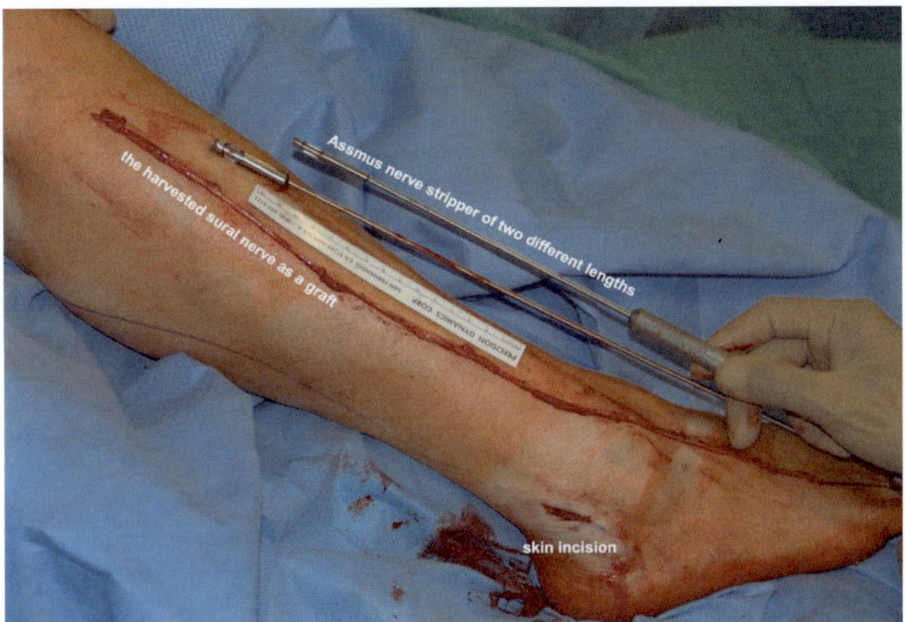

Fig. 5.7 An intraoperative photograph of harvesting the entire sural nerve through a 1 inch skin incision behind the lateral malleolus using the Assmus nerve stripper

introduced [21]. Albeit being minimally invasive, the endoscopic nerve harvest is quite time consuming. A much simpler and minimally invasive option was described by one of the editors of this book (HA): the use of a nerve stripper [1]. This is the preferred technique of the author of this chapter [11]. The technique is as follows [referring to the sural nerve—the most harvested nerve for reconstruction]: the sural nerve is exposed at the level of the lateral malleolus and transected here. With the use of the Assmus nerve stripper, the entire length of the nerve up to its origin from the popliteal fossa is dissected and stripped off (Fig. 5.7). The entire procedure takes approximately 10–20 min depending on the proficiency of the surgeon. One technical disadvantage of this method is the harvest of only a part of the sural nerve, especially in the eventuality of the anatomical variation that the nerve bi- or trifurcates quite proximally. In this case, a resistance is felt by the surgeon at the point of bifurcation (usually half way up the dorsal aspect of the calf), and a second skin incision there might become necessary.

The one main critique the stripping technique has brought forth is the shearing forces acting on the graft as the nerve is being manipulated. However, it has been elegantly demonstrated by a microscopic study that the stripped nerves are no more damaged than those that had been harvested in the open manner [7]. Furthermore, as already mentioned, as we graft a nerve, we do not transplant the nerve fascicles or axons; essentially we transplant the Schwann cell scaffolding, which can impossibly be damaged by manipulations no graver than severe homogenization.

5.4 Techniques in Intraoperative Diagnostics

There are two kinds of intraoperative diagnostic tools in peripheral nerve lesions, viz., that of form and that of function. In former times, the fibrotic nerve was palpated so that the surgeon could "feel" its consistency. When found necessary, the nerve sheath was opened, and the fascicles were explored under the magnification of the operating microscope. The latter manipulation does run the risk of additional iatrogenic fibrosis. In modern times, intraoperative *ultrasonographic imaging* is able to offer much information about the continuity and condition of the nerve fascicles without having to open the nerve sheath. Neurosonography has become a standard diagnostic tool for not only judging the grade of fibrosis but also to detect occult lesions along the course of the explored nerve that otherwise would go undetected. Intraoperative neurosonography is also an important tool to control the extent of resection in some types of tumors affecting or encompassing peripheral nerves [6, 19].

A nerve lesion that shows structural integrity does not necessarily mean that the conduction across the lesion is intact. Intraoperative electrophysiological studies have come to play a significant role in modern peripheral nerve surgery [3, 16]. Intraoperative electrophysiology can be subdivided into two categories: (a) "monitoring" an intact nerve function during a manipulation inside the nerve, e.g., whilst removing an intraneural tumor, and (b) "diagnosing" the functional integrity of the lesioned fascicles of a peripheral nerve. For monitoring, somatosensory evoked potential (SSEP) and motor evoked potential (MEP) tests are the two salient methods. In addition to this, my preferred technique is to record SSEP from the head leads and electromyographic potentials (EMG) from the target muscle supplied by the motor nerve while directly stimulating the nerve fascicles during tumor removal. Specifically designed stimulation-integrated microdissectors aid in performing the microsurgical steps of tumor resection and simultaneously stimulate the fascicles. As opposed to monitoring and preserving the integrity of conducting nerve fascicles during an intraneural manipulation, the method of choice for intraoperative diagnosis is to measure the nerve conduction velocity by means of recording the compound nerve action potential (cNAP) across a nerve lesion. In this method, the nerve is exposed both proximal and distal to the lesion it lodges; the electrical stimulation is applied to the nerve proximal to the lesion with a tripolar stimulation electrode in the shape of a trident hook, and the cNAP is recorded using a bipolar hook electrode lead placed distal to the nerve lesion (Fig. 5.8). Sometimes it is possible to recognize partial lesions of nerves and replace only those fascicles that do not show conduction—effectuating the so-called split repair.

Surgical techniques in peripheral nerve lesions have evolved rapidly with the availability of technological advancement. The technological breakthrough in many areas of surgery notwithstanding the basic principles of peripheral nerve surgery will continue to remain the same. These are as follows: (a) expose the nerve adequately proximal and distal to the area of lesion; (b) inspect, examine, and study the lesion meticulously from both the morphological and functional points of view; (c) constantly endeavor to preserve any available function of the nerve fascicles

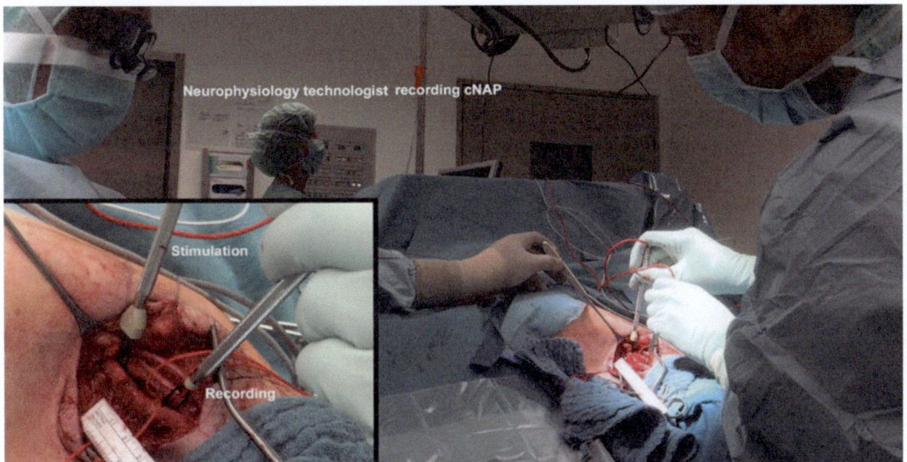

Fig. 5.8 Intraoperative recording of compound nerve action potential. Inset shows the nerve being stimulated using a tripolar trident hook electrode proximal to the lesion and the potentials captured using a bipolar hook electrode. The technologist is seen in the background

while treating a nerve lesion; and lastly (d) treat only those fascicles that are non-functional with a hope to render them functional.

The means to achieve the abovementioned goals will obviously vary with two basic qualities of the future nerve surgeon, viz., his open-mindedness to apply new techniques and his familiarity with evolving technology in every walk of our fascinating times.

References

1. Assmus H. Sural nerve removal using a nerve stripper. Neurochirurgia (Stuttg). 1983;26:51–2.
2. Barton MJ, Morley JW, Stoodley MA, Lauto A, Mahns DA. Nerve repair: toward a sutureless approach. Neurosurg Rev. 2014;37(4):585–95.
3. Campbell WW. Evaluation and management of peripheral nerve injury. Clin Neurophysiol. 2008;119(9):1951–65.
4. Dützmann S, Martin KD, Sobottka S, Marquardt G, Schackert G, Seifert V, Krishnan KG. Open vs retractor-endoscopic in situ decompression of the ulnar nerve in cubital tunnel syndrome: a retrospective cohort study. Neurosurgery. 2013;72(4):605–16.
5. Hansen TB, Majeed HG. Endoscopic carpal tunnel release. Hand Clin. 2014;30(1):47–53.
6. Koenig RW, Schmidt TE, Heinen CP, Wirtz CR, Kretschmer T, Antoniadis G, Pedro MT. Intraoperative high-resolution ultrasound: a new technique in the management of peripheral nerve disorders. J Neurosurg. 2011;114(2):514–21.
7. Koller R, Frey M, Rab M, Deutinger M, Freilinger G. Which changes occur in nerve grafts harvested with a nerve stripper? Morphological studies. Handchir Mikrochir Plast Chir. 1995;27(2):98–101. German.
8. Krishnan KG, Schoeller K, Uhl E. The application of a compact HD-exoscope for illumination and magnification in high-precision surgical procedures. World J Neurosurg. Accepted 2016.
9. Krishnan KG. Endoscopic decompression of the tarsal tunnel. Tech Foot Ankle Surg. 2010;9(2):52–7.

10. Krishnan KG. Part 4: appendices. In: Krishnan KG, editor. A handbook of flap raising techniques. Stuttgart/New York: Georg Thieme Verlag; 2008. p. 93–106.
11. Krishnan KG. Harvesting of the sural nerve using a stripper. Neurosurgery. 2007;60(1):E208.
12. Krishnan KG, Pinzer T, Schackert G. A novel endoscopic technique in treating single nerve entrapment syndromes with special attention to ulnar nerve transposition and tarsal tunnel release – clinical application. Neurosurgery. 2006;59(1):ONS 89–100.
13. Krishnan KG, Pinzer T, Reber F, Schackert G. The endoscopic exploration of the brachial plexus: technique and topographical anatomy. Neurosurgery. 2004;54(2):401–9.
14. Kriss TC, Kriss VM. History of the operating microscope: from magnifying glass to microneurosurgery. Neurosurgery. 1998;42(4):899–908.
15. Litynski GS. Endoscopic surgery: the history, the pioneers. World J Surg. 1999;23(8): 745–53.
16. Malessy MJ, Pondaag W, van Dijk JG. Electromyography, nerve action potential, and compound motor action potentials in obstetric brachial plexus lesions: validation in the absence of a "gold standard". Neurosurgery. 2009;65(4 Suppl):A153–9.
17. Mamelak AN, Danielpour M, Black KL, Hagike M, Berci G. A high-definition exoscope system for neurosurgery and other microsurgical disciplines: preliminary report. Surg Innov. 2008;15(1):38–46.
18. Menovsky T, Beek JF. Laser, fibrin glue, or suture repair of peripheral nerves: a comparative functional, histological, and morphometric study in the rat sciatic nerve. J Neurosurg. 2001;95(4):694–9.
19. Pedro MT, Antoniadis G, Scheuerle A, Pham M, Wirtz CR, Koenig RW. Intraoperative high-resolution ultrasound and contrast-enhanced ultrasound of peripheral nerve tumors and tumor-like lesions. Neurosurg Focus. 2015;39(3):E5.
20. Shirzadi A, Mukherjee D, Drazin DG, et al. Use of the Video Telescope Operating Monitor (VITOM) as an alternative to the operating microscope in spine surgery. Spine (Phila Pa 1976). 2012;37(24):E1517–23.
21. Spinks TJ, Adelson PD. Pediatric sural nerve harvest: a fully endoscopic technique. Neurosurgery. 2009;64(5):360–3; discussion 363–4.

Specific Challenges in Brachial Plexus Surgery

6

Thomas J. Wilson and Lynda J.-S. Yang

Contents

6.1 Introduction

Brachial plexus injuries are devastating, resulting in loss of function of the upper extremity, which carries significant morbidity. In adults, trauma is the most common etiology of brachial plexus injury. In neonates, the exact pathophysiology of brachial plexus injuries is unclear but occurs before or during labor and parturition [1]. Neonatal brachial plexus palsy (NBPP) occurs in approximately 1 in 1000 live births [5]. A significant proportion of these patients will demonstrate spontaneous recovery with therapy alone and no operative intervention. However, there remains a subset of these patients that will not recover without operative intervention.

Until only recently, adult and neonatal brachial plexus palsies were thought of as nonsurgical pathologies. Little was available in the way of surgical treatment. Early efforts had poor results which discouraged continuing surgical treatment [32]. World

T.J. Wilson
Department of Neurosurgery, Stanford University, Palo Alto, CA, USA
e-mail: thowil@med.umich.edu

L.J.-S. Yang, MD, PhD (✉)
Department of Neurosurgery, University of Michigan, Ann Arbor, MI, USA
e-mail: ljsyang@med.umich.edu

© Springer International Publishing AG 2017
K. Haastert-Talini et al. (eds.), *Modern Concepts of Peripheral Nerve Repair*,
DOI 10.1007/978-3-319-52319-4_6

War II ultimately revived the interest in repair of adult brachial plexus injuries, and during this time, Seddon pursued repair with improved outcomes, sparking a renewed interest. Neonatal brachial plexus palsy, however, remained a nonsurgical condition until the work of Gilbert revived interest when he reported improved outcomes and, in particular, improved safety of operative intervention [27, 29]. As surgery has increasingly become an option and new innovative techniques have been employed, a number of challenges have arisen that span the gamut from preoperative evaluation and decision-making to intraoperative decisions regarding the optimal nerve reconstruction strategy to be performed to evaluating outcomes in these patients postoperatively. In this chapter, we highlight specific challenges facing the peripheral nerve surgeon in each phase of care and highlight the areas needing further research. While the majority of these specific challenges pertain to the NBPP population, decisions regarding whether to perform nerve graft repair or nerve transfer pertain to both the NBPP population and adult population, and both will be highlighted. As research continues and new innovative techniques for evaluation and treatment are developed, these specific challenges are likely to be overcome, but with progress, new challenges and new questions are likely to be raised.

6.2 Challenges in the Preoperative Evaluation

Preoperatively, the main challenges facing the peripheral nerve surgeon when evaluating a patient with NBPP are (1) determining whether or not to operate and (2) the optimal timing of operative intervention. This begs the question, what is the optimal method of evaluation to guide this decision-making? While a significant proportion of patients with NBPP will recover spontaneously if given time, data also have shown that earlier operative intervention is associated with improved outcomes following graft repair or nerve transfer [8, 36]. Thus, early dichotomization of patients into those likely to spontaneously recover and those unlikely to spontaneously recover has great importance. The most fundamental question to be addressed by all methods of evaluation that informs the likelihood of recovery is: what is the nature of the injury? Lesions likely to recover include neurapraxic injuries and axonotmetic injuries. Those lesions with no hope of spontaneous recovery include nerve root avulsions (preganglionic) and postganglionic, neurotmetic lesions (ruptures).

The mainstay of evaluation of these patients remains the physical examination. While documenting a baseline examination shortly after birth is extremely important, little is gleaned with regard to prognostication from this initial examination. The exception may be the presence of Horner's syndrome in the context of a panplexopathy. The presence of Horner's syndrome is indicative of a preganglionic, non-recoverable lesion and an indication for surgery [3]. Aside from this finding, there are no reliable indicators of non-recoverable lesions. Hence, time must be allowed to observe for spontaneous recovery. Though the optimal time period is not universally agreed upon, the most commonly used time period is 3 months. Gilbert demonstrated that motor outcomes at 5 years of age were poor in those children who failed to spontaneously recover biceps function by 3 months of age [23, 27, 28].

Thus, this is the rationale for evaluation at 3 months, with those children not demonstrating spontaneous recovery of biceps function being unlikely to recover and thus likely to benefit from operative intervention.

However, further detailed analysis revealed flaws in this system. Michelow and colleagues demonstrated that if absent biceps function at 3 months is utilized as the sole criterion for prediction of recovery, the prediction is incorrect in 12% of patients. When multiple movements were assessed at 3 months and combined into an overall score, the percentage of incorrect predictions dropped to only 5% [7]. One of the issues with assessment at 3 months of age is that some patients will go on to develop biceps contraction at 6 months, though the significance of this finding is uncertain [39, 44]. Waters has shown that patients developing biceps function after 5 months of age have improved outcomes with operative management compared to nonoperative management [8, 18]. Thus, the significance of delayed recovery of biceps function is unclear. Other tests such as the towel test and cookie test have been suggested to be helpful in predicting those patients likely to benefit from surgery [9, 13, 38]. In the towel test, a towel is placed over the infant's face and the infant is observed for the ability to remove the towel with the affected arm [9]. In the cookie test, a small cookie is placed in the infant's hand and the humerus is held at the infant's side. The infant is then observed for the ability to generate enough elbow flexion to bring the cookie into his/her mouth [13]. This remains a specific challenge to the peripheral nerve surgeon as there is no consensus as to what method of evaluation should be used. The ideal evaluation would be highly specific and sensitive and able to be predictive at a young age.

In adults, one of the mainstays of evaluation of the peripheral nervous system is the electrodiagnostic study including nerve conduction studies and electromyography (EMG), but these studies are fraught with difficulties in neonates. EMG studies are often difficult to interpret and are often discordant with clinical findings. When a paralyzed biceps is encountered clinically, one would expect an EMG to show a loss of motor unit potentials (MUPs) and the presence of denervation activity. However, frequently, in the setting of a paralyzed biceps in infants, motor unit potentials are present and denervation activity is absent [36]. A number of reasons have been suggested for these confusing findings. Malessy and colleagues have suggested five reasons that there may be the presence of motor unit potentials despite no observed biceps activity: (1) inadequacy of the clinical examination, (2) overestimation of the number of motor unit potentials, (3) luxury innervation, (4) central motor disorders, and (5) abnormal nerve branching [11]. Examining an infant is limited by the inability of the infant to voluntarily participate in the examination. For this reason, it may be that Medical Research Council (MRC) grade 1 or 2 movement may be missed. The estimate of the number of motor unit potentials (MUPs) may be overestimated due to the difference in the size of motor fibers in infants versus adults. Because fibers are smaller in infants, a significantly larger number of fibers are recorded for the same EMG needle uptake area compared to adults. Luxury innervation refers to the idea that muscles have more than one neuromuscular synapse early in development. During normal development, pruning occurs so that only one neuromuscular synapse remains. However, there is disagreement about when this pruning occurs. If this pruning occurs after birth, it may be that the presence of a brachial plexus lesion affects this pruning process. It has previously been shown that in infants with NBPP, intraoperative stimulation of C7 yields elbow flexion and shoulder abduction, suggesting luxury innervation of the biceps by C7 [10, 40]. This luxury innervation is not pruned due to the lack of competition from C5 and C6 as a result of the brachial plexus injury. This may result in identifying MUPs in the biceps from C7 rather than C5 or C6. Many motor pathways depend on afferent input for normal formation. However, in NBPP, not only is the motor pathway lost but the afferent sensory pathway is also lost. This may result in abnormal formation of central motor pathways such that even if axonal regeneration occurs to the biceps, the motor pathways may not form correctly to allow movement [30, 51]. Finally, abnormal branching of regenerating axons may occur. Because of abnormal branching and misdirection, axonal regeneration can terminate in other muscles resulting in co-contraction of various muscles. This co-contraction due to abnormal branching may result in detection of MUPs despite lack of activation of the biceps.

With all of the incumbent challenges of EMG and nerve conduction studies in neonates, we are left to ask whether or not there is any value to obtaining such studies. There does still appear to be some value to obtaining these studies, and we still do routinely obtain them. Electrodiagnostic studies can be poor at detecting nerve root avulsions. We have previously shown that the sensitivity for nerve root avulsions is only 27.8%. However, electrodiagnostic studies do appear to be useful in detecting ruptures. The sensitivity of electrodiagnostic studies for intraoperatively confirmed ruptures was 92.8%. This pattern is the opposite pattern compared to computed tomographic (CT) myelography which showed increased sensitivity for avulsions and lower sensitivity for ruptures. Thus, the two studies complement each other [24]. Electrodiagnostic studies do potentially provide useful information, though their interpretation and optimal timing remain challenges in the evaluation of NBPP.

In addition to the clinical examination and electrodiagnostic studies, a variety of imaging modalities are available to aid in the evaluation of the patient with NBPP. However, each modality comes with its own set of challenges, and no consensus exists for the appropriate set of diagnostic imaging for these patients. Historically, CT myelography is likely the most commonly employed imaging modality in these patients. We have shown previously that CT myelography has only a 58.3% sensitivity for nerve ruptures but a 72.2% sensitivity for avulsions [24]. While this adds valuable information, CT myelography is certainly not highly sensitive. Debate also exists as to what criteria should be used to diagnose an avulsion. The two most debated criteria are pseudomeningocele alone versus pseudomeningocele with absent rootlets. Studies vary in the reported value of each of these diagnostic criteria. Tse et al. reported a sensitivity of 73% when pseudomeningocele alone was used versus 68% when pseudomeningocele with absent rootlets was used. While not highly sensitive, CT myelography is highly specific with reported specificity of 96% whether pseudomeningocele alone or with absent rootlets was used [21]. A previous report from Chow and colleagues had shown that utilizing pseudomeningocele with absent rootlets for diagnosis improved the specificity from 85% to 98% [12]. One possible explanation for why Tse and colleagues did not find a similar increase is that their cohort of patients had a high proportion of Narakas grade 3 and 4 injuries and thus included more injuries to C8 and T1 where avulsions are more likely to occur. In their study, 18 of 19 pseudomeningoceles identified contained absent rootlets. If they had had a more mixed population relative to injury severity and level, they may have observed a similar increase in specificity as Chow observed [21]. Regardless, the optimal diagnostic criteria remain debated and sensitivity remains a challenge. Additionally, CT myelography brings with it challenges inherent to the procedure including the invasive nature of the procedure, instillation of intrathecal contrast and associated risks, and exposure to ionizing radiation.

More recently, high-resolution magnetic resonance (MR) imaging and MR myelography have been used in place of and compared to CT myelography. MR myelography has been shown to have a similar sensitivity and specificity for nerve root avulsions compared to CT myelography, 68% and 96%, respectively [21]. In another study of high-resolution MR imaging, the sensitivity and specificity for nerve root avulsions were 75% and 82%, respectively [48]. Some of the same issues are present as with CT myelography, however, including defining the diagnostic criteria to be used for avulsions and imaging of the more distal nerves for evidence of rupture. High-resolution MR imaging/MR myelography does offer some advantages, including the noninvasive nature of the study, the lack of intrathecal contrast administration, and the lack of exposure to ionizing radiation. With a similar sensitivity and specificity compared to CT myelography and the several aforementioned advantages, we have replaced CT myelography with high-resolution MR imaging in the evaluation of patients with NBPP.

One challenge of both CT and MR myelography is visualization of the extraforaminal nerve roots and trunks in order to evaluate for evidence of rupture. Ultrasound can help overcome this challenge. Ultrasound is particularly useful in evaluating the upper and middle trunks and less so the lower trunk. The sensitivity in identification of neuromas in our study was 84% for both the upper and middle trunks and 68% for the lower trunk. Ultrasound can also be used to provide some information about how proximal the injury is based on evaluation of the serratus anterior and rhomboid muscles. Atrophy in these muscles detected on ultrasound suggests a proximal injury, making the presence of a viable proximal stump for nerve grafting less likely and making us favor nerve transfer instead [25]. Ultrasound has little ability to evaluate the preganglionic segments of nerve roots, making evaluation for avulsion difficult with this imaging modality.

One of the main challenges in preoperative decision-making is identification of appropriate candidates for nerve surgery as early as possible. To that end, we attempted to identify peripartum and neonatal factors that were associated with persistent NBPP. We identified cephalic presentation, induction or augmentation of labor, birth weight > 9 lbs., and the presence of Horner's syndrome as increasing the likelihood of persistence. Cesarean delivery and Narakas grade 1 and 2 injuries reduced the odds of persistence. Horner's syndrome is a constellation of clinical findings including ptosis, anhidrosis, and miosis due to injury to the sympathetic trunk. The Narakas scale is an injury grading scale where Narakas grade 1 is injury to the upper trunk only (C5, C6), grade 2 is injury to the upper and middle trunks (C5, C6, C7), grade 3 is pan-plexus injury without Horner's syndrome, and grade 4 is a pan-plexus injury with Horner's syndrome. The study was performed on a biased sample, however, due to the fact that the population of patients was that already referred for evaluation by a nerve surgeon [37]. Nonetheless, the design of this study was such that it was intended to address the main challenge in the preoperative evaluation of patients with NBPP which is early identification of those patients that will not recover who should undergo nerve surgery. While the physical examination, electrodiagnostic studies, imaging studies, and peripartum/neonatal history all have a role in the evaluation, we are still in need of a predictive algorithm that incorporates all of these methods of evaluation that can dichotomize these patients with high sensitivity and specificity. Future research should continue to address this challenge. Until such research addresses this challenge, we have developed our own algorithm for evaluation at the University of Michigan (Fig. 6.1).

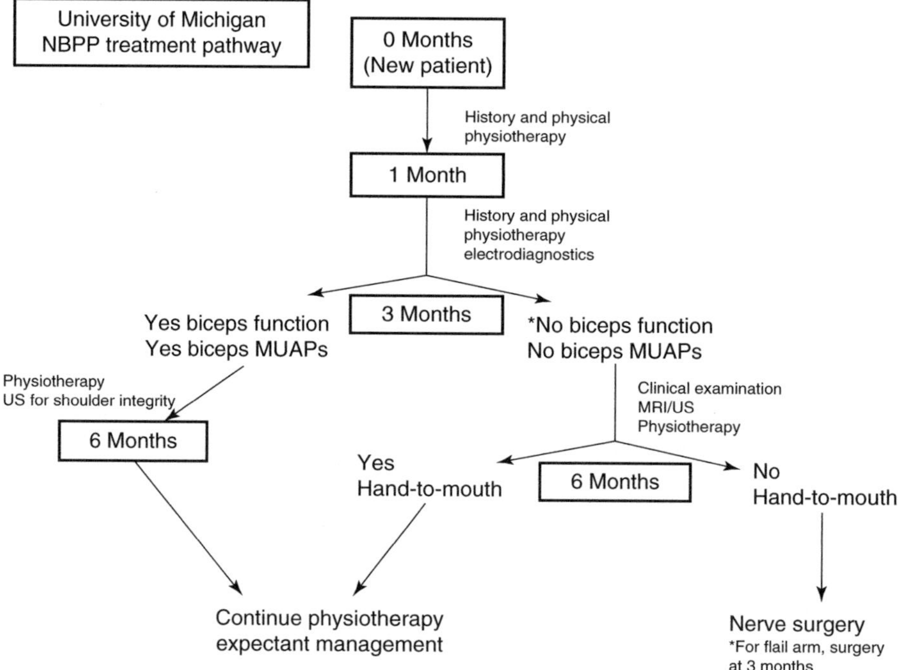

Fig. 6.1 Flowchart of the University of Michigan Neonatal Brachial Plexus Palsy (NBPP) pathway of presurgery decision-making. *US* Ultrasound, *MRI* Magnet Resonance Imaging, *MUAP* Motor Unit Action Potential

6.3   Challenges in Intraoperative Decision-Making

Once a decision is made to operate on a patient for persistent NBPP, a number of intraoperative challenges face the nerve surgeon. The main decision is what intervention to perform: neurolysis alone, nerve graft repair, or nerve transfer. For a number of reasons, this decision remains challenging. One main reason is the lack of comprehensive postoperative data that allow head-to-head comparison of interventions. This will be discussed in the next section. With the available data, how does the nerve surgeon make this decision?

In adults, recoding nerve action potentials (NAPs) across a lesion in continuity can be helpful. When nerve action potentials are recorded across a lesion, it is often best to perform neurolysis alone, as nerve action potentials traveling across a lesion in continuity suggest a recovering nerve [43]. However, in neonates, nerve action potentials are not similarly useful. Intraoperative nerve action potentials in neonates are thought to provide overly optimistic data. One study included ten lesions in continuity and found positive NAPs across the lesion in five patients. Neurolysis alone was performed in these patients and none had a good recovery [33]. In an additional study, Pondaag and colleagues found that the specificity for a severe lesion of absent NAPs and compound muscle action potentials (CMAPs) across a lesion in continuity was high (>90%). However, the sensitivity was very low (<30%) [41]. Taken together, the available data suggest that intraoperative NAPs and CMAPs in neonates are not useful in guiding decisions. Thus, the surgeon is challenged with relying on preoperative assessment to determine who should undergo nerve reconstruction and that is fraught with the challenges previously described.

Thus, once a decision for surgery is made, the real decision is whether to graft or to transfer. There are very little data and very few studies directly comparing nerve grafts to nerve transfers for NBPP. Thus, determining the optimal intervention remains challenging. There are currently disagreements about the role of nerve transfers in the treatment of NBPP. The International Federation of Societies for Surgery of the Hand suggests that the role of nerve transfers in NBPP is unclear but that nerve transfers are a viable option for Erb's palsy but should not be first-line treatment for more severe injuries. The committee suggests that there should not be an overreliance on nerve transfers and there should remain an inclination toward brachial plexus exploration and nerve graft repair [52]. Further data, however, are needed to determine the optimal roles of both nerve transfer and nerve graft repair.

Erb's palsy with C5 and C6 injury is the most common pattern of injury in NBPP. While nerve graft repair is the traditional intervention, nerve transfers have been shown to be a viable option. Recovery of elbow flexion has been shown to be good following ulnar or median nerve fascicle transfer to the biceps or brachialis branch of the musculocutaneous nerve. In one study, 87% of patients undergoing these transfers obtained functional elbow flexion recovery. Outcomes were worse for supination recovery with only 21% recovering functional supination [34]. While there was no direct comparison to nerve graft repair, these outcomes suggest nerve transfer is a viable option.

Reinnervation of the suprascapular nerve is important for restoration of external rotation of the shoulder following C5/C6 injury in NBPP. Early experience reinnervating the suprascapular nerve was poor regardless of whether nerve graft repair or nerve transfer was used [35]. More recently, however, outcomes have been better. There have been mixed data comparing spinal accessory nerve transfer with C5 nerve graft repair. Spinal accessory nerve transfer is at least equivalent to C5 nerve graft repair, but some data suggest it may have better outcomes [47, 53]. Seruya and colleagues found that C5 nerve graft repair led to poorer shoulder function and also increased secondary shoulder surgery compared to spinal accessory to suprascapular nerve transfer [47]. The major challenge remains making a decision to graft or to transfer in the setting of a lack of data comparing the two interventions. Future studies will need to focus on comparing outcomes. Additionally, as we discuss in the next section, it will be important to compare outcomes more in depth than simply motor outcome.

A similar dilemma exists in the adult population of brachial plexus injury patients. What is the optimal repair strategy to maximize outcomes? For upper trunk injuries with loss of shoulder abduction, external rotation, and elbow flexion, there is little in the way of direct comparisons between nerve graft repair and nerve transfer. However, two recent meta-analyses help compare the two strategies, and both concluded that nerve transfer strategies are superior to nerve graft repair. These studies utilized the Medical Research Council (MRC) grading scale where M5 is normal strength, M4 is movement against active resistance, M3 is movement against gravity but no active resistance, M2 is movement with gravity eliminated, and M1 is flicker movement or contraction only. Garg and colleagues found that 83% of patients with nerve transfers achieved M4 or greater elbow flexion strength and 96% achieved M3 or greater. Comparatively, only 56% of patients with nerve graft repair achieved M4 or greater strength and 82% achieved M3 or greater. Shoulder outcomes were similarly better with nerve transfers. Seventy-four percent of dual nerve transfer patients achieved M4 or greater shoulder abduction strength versus 46% with nerve graft repair. Both shoulder abduction and external rotation were better in the nerve transfer group [26]. Ali and colleagues recently supported these findings. They found that nerve transfer techniques were superior to nerve graft repair for the restoration of elbow flexion and shoulder abduction. Specifically, with regard to elbow flexion, the Oberlin procedure (transfer of an ulnar fascicle to the biceps branch of the musculocutaneous nerve) was superior to all other strategies [4]. Thus, for upper trunk brachial plexus injuries, nerve transfer seems to be superior to nerve graft repair, but no direct comparative data are available. This data is not conclusive, however, and there certainly remains controversy. In fact, in a systematic review, we previously found that the data did not support the sole use of nerve transfers for upper brachial plexus injury. We recommended at that time that the standard should still include brachial plexus exploration with nerve graft repair when feasible [55]. Additional comparative studies are needed to better elucidate the optimal strategy.

Restoration of hand function following lower trunk injuries is similarly challenging. In addition to nerve graft and nerve transfer techniques, an additional consideration is the Doi procedure (double free muscle transfer) [20]. Ray and colleagues initially described a series of four patients with isolated lower trunk injuries in whom they performed transfer of the nerve to the brachialis to the anterior interosseous nerve, with good clinical outcomes [42]. Isolated lower trunk injuries, however, are relatively uncommon. With concomitant involvement of the upper brachial plexus, nerve transfer options become more limited. Dodakundi and colleagues initially reported success of the double free muscle transfer in total brachial plexus injury [19]. As an adjunctive intervention, wrist arthrodesis has been shown to improve both finger range of motion and overall hand function in patients with double free muscle transfer for pan-plexus injury [2]. Recently, Satbhai and colleagues reported an improvement in overall functional outcome and quality of life using the double free muscle transfer versus single free muscle transfer or nerve transfer for patients with pan-plexus injury [46]. However, it is not clear that hand function was significantly better. In addition, this study pertains to patients with pan-plexus injury and focuses on the overall function of the limb. In cases of isolated lower trunk injury, it is not clear what strategy, whether nerve graft, nerve transfer, free muscle transfer, or tendon transfer, yields the best results. Thus, determining the optimal reconstructive strategy remains challenging.

6.4 Challenges in Postoperative Evaluation

Postoperatively or, in the case of those neonates who are managed nonoperatively, throughout the natural history of the condition, we are tasked with evaluating these children in some way. This is particularly important in order to collect data to determine if operative intervention is helpful and in order to compare different types of intervention head to head. To this point, most evaluations have focused on motor outcomes and grading individual motor movements on scales such as the Medical Research Council (MRC), Active Movement Scale (AMS), and Louisiana State University motor grading scales. While a variety of outcome measures have been used, the five most common in the published literature include range of motion of the shoulder, range of motion of the elbow, the Mallet scale, MR imaging findings, and the MRC grading scale [45]. Very few evaluation instruments/metrics are specifically validated for use in the NBPP population. Validated evaluation instruments/metrics include the Active Movement Scale, Toronto Scale Score, Mallet Score, Assisting Hand Assessment, and Pediatric Outcomes Data Collection Instrument [16]. While gross motor function and evaluation of body structure and function are important, this may not capture the complete picture, as simply grading motor strength ignores other important factors such as sensation, arm preference, proprioception, functional use of the extremity, cognitive development, pain, quality of life, and language development [22]. Thus, it remains a specific challenge to determine how best to evaluate patients with NBPP. While a number of these domains of evaluation are specifically to the NBPP population, a similar problem exists when evaluating adults with brachial plexus injury following intervention. In this population, it also remains a specific challenge to go beyond purely evaluating motor recovery and rather to also evaluate quality of life, functional use of the affected limb, and pain [22].

One challenge of the postoperative evaluation is determining the optimal duration of time to follow these patients. From age 5 onward, these patients generally have stable to improved hand and shoulder function. However, over the same time course, elbow function tends to slightly deteriorate. This is true whether or not nerve reconstruction was performed. Children who have poor shoulder external rotation benefit from shoulder surgery with significant improvement postoperatively [50]. Because of the continued decrease in elbow function and the significant benefit to shoulder external rotation following surgery for those patients in whom external rotation limitation is recognized, it is important to follow these patients throughout childhood and adolescence and into adulthood.

In the general population, approximately 90% of people have a right arm preference/dominance. In children with left upper extremity brachial plexus palsy, that percentage remains roughly the same, 93% in our previous study. However, when the right upper extremity is the affected limb, only 17% preferred the right limb. This is a significant deviation away from the population average [54]. This suggests neural plasticity is at work early in the development of these children. However, what is not clear is how dominant the unaffected extremity becomes. Is the affected extremity essentially a useless limb, or is there only a slight preference for the unaffected extremity? More importantly, do surgical interventions improve the functional use of the extremity and reduce the preference for the unaffected extremity? Finally, do nerve transfers that offer earlier, though some would argue less complete, recovery offer advantages over nerve graft repair due to the fact that recovery occurs when motor patterns are being established? These are the challenges in evaluation that remain to be answered.

It may not simply be weakness that leads to altered limb preference and reduced functionality. Proprioception plays a large role in the functional use of extremities. However, to this point, little focus has been given to evaluating proprioception following brachial plexus injury. We have previously assessed elbow position sense in adolescents with a history of NBPP. We found that position sense is impaired in the affected limb following NBPP [14]. Similarly, tactile spatial perception is reduced in the hand of the affected limb following NBPP [15]. It is unclear how much this affects daily use of the limb and overall limb preference. However, it may be an important component not assessed by purely focusing on gross motor function. Further assessments of proprioception and advanced sensory modalities are needed in future studies to determine their importance in daily activities and which interventions improve these modalities that contribute to complex functional use.

Delayed or altered use of the affected limb may also affect development in a more global fashion. Motor impairments in children have previously been reported to delay language [31]. The nature of the relationship between motor function and language is unclear. Decreased motor function may impair the ability of the child to explore the world around them, thus delaying language. We have previously shown a high rate of language delay in toddlers with a history of NBPP [17]. This finding has several important implications. First, it suggests that treating children with

NBPP is more complex than simply focusing on motor rehab. Recognizing the association of language delay and NBPP means that rehabilitation focused on language development should be part of the overall rehabilitation program. Furthermore, it suggests that assessment of language is an important component of the global assessment of these patients. A further understanding of exactly how language development and motor deficits, and more specifically NBPP, are linked may lead to a better understanding of interventions that may address this issue. For example, if delays in language development result from a decreased ability to explore the surroundings at a very young age, those interventions that favor early recovery, i.e., nerve transfers as opposed to nerve graft repair, may favor improved language development. This remains hypothetical, however, but points to the challenge of needing more complex evaluations to determine optimal interventions.

With language development being affected, one might hypothesize that behavioral issues may arise in children with a history of NBPP. This hypothesis turns out to be correct. Children with a history of NBPP show global developmental delays, difficulty with hand-eye coordination, and a higher incidence of emotional and behavioral problems. This was closely associated with the severity of initial injury [6]. One might assume that earlier or more complete recovery may be associated with a reduction in behavioral problems, but this has never been demonstrated. Thus, it remains a challenge to evaluate behavioral outcomes and to determine what factors are associated with reduced behavioral issues, including which interventions may help reduce these issues.

All of these challenges point to need for more global and comprehensive evaluation of patients with NBPP, both managed operatively and nonoperatively. Ultimately, what is important to these patients is having the highest quality of life possible. A number of factors have been identified as affecting the quality of life in these patients including social impact and peer acceptance, emotional adjustment, aesthetics and body image, functional limitations, finances, pain, and family dynamics [49]. The diversity of these factors points to the fact that assessment necessarily involves more than simply assessing motor function. It remains the challenge of the nerve surgeon taking care of patients with NBPP to develop the optimal assessment metrics and intervals and to compare interventions head to head using optimized global metrics, ultimately moving beyond simply the World Health Organization International Classification of Functioning, Disability, and Health Body Function and Structure domain and moving into evaluations in the Activity and Participation domain (http://www.who.int/classifications/icf/en/).

Conclusion

Neonatal brachial plexus palsy is a relatively common pathology. While most children will recover without surgical intervention, a number of challenges face the nerve surgeon throughout the preoperative, intraoperative, and postoperative care of these patients. Similar dilemmas regarding nerve graft repair versus nerve transfer face both the nerve surgeon treating NBPP and adult brachial plexus injury. Surgery for NBPP is in its relative infancy, which is the origin of most of these challenges. Further data are needed to help overcome these obstacles and guide decision-making for these patients. While these challenges remain, it is an exciting field that holds promise for helping to improve function and quality of life for these patients through progressively improved decision-making algorithms and surgical intervention. With progress, however, new questions are likely to arise that will continue to challenge the nerve surgeon in optimizing care of these patients.

References

1. Executive summary: neonatal brachial plexus palsy. Report of the American College of Obstetricians and Gynecologists' Task Force on Neonatal Brachial Plexus Palsy. Obstet Gynecol. 2014;123:902–4.
2. Addosooki A, Doi K, Hattori Y, Wahegaonkar A. Role of wrist arthrodesis in patients receiving double free muscle transfers for reconstruction following complete brachial plexus paralysis. J Hand Surg Am. 2012;37:277–81.
3. Al-Qattan MM, Clarke HM, Curtis CG. The prognostic value of concurrent Horner's syndrome in total obstetric brachial plexus injury. J Hand Surg Br. 2000;25:166–7.
4. Ali ZS, Heuer GG, Faught RW, Kaneriya SH, Sheikh UA, Syed IS, et al. Upper brachial plexus injury in adults: comparative effectiveness of different repair techniques. J Neurosurg. 2015;122:195–201.
5. Bager B. Perinatally acquired brachial plexus palsy – a persisting challenge. Acta Paediatr. 1997;86:1214–9.
6. Bellew M, Kay SP, Webb F, Ward A. Developmental and behavioural outcome in obstetric brachial plexus palsy. J Hand Surg Br. 2000;25:49–51.
7. Bertelli JA, Ghizoni MF. Nerve transfer from triceps medial head and anconeus to deltoid for axillary nerve palsy. J Hand Surg Am. 2014;39:940–7.
8. Bertelli JA, Ghizoni MF. Reconstruction of C5 and C6 brachial plexus avulsion injury by multiple nerve transfers: spinal accessory to suprascapular, ulnar fascicles to biceps branch, and triceps long or lateral head branch to axillary nerve. J Hand Surg Am. 2004;29:131–9.
9. Bertelli JA, Ghizoni MF. The towel test: a useful technique for the clinical and electromyographic evaluation of obstetric brachial plexus palsy. J Hand Surg Br. 2004;29:155–8.
10. Bertelli JA, Ghizoni MF. Transfer of the accessory nerve to the suprascapular nerve in brachial plexus reconstruction. J Hand Surg Am. 2007;32:989–98.
11. Bertelli JA, Kechele PR, Santos MA, Duarte H, Ghizoni MF. Axillary nerve repair by triceps motor branch transfer through an axillary access: anatomical basis and clinical results. J Neurosurg. 2007;107:370–7.
12. Bertelli JA, Tacca CP, Winkelmann Duarte EC, Ghizoni MF, Duarte H. Transfer of axillary nerve branches to reconstruct elbow extension in tetraplegics: a laboratory investigation of surgical feasibility. Microsurgery. 2011;31:376–81.
13. Borschel GH, Clarke HM. Obstetrical brachial plexus palsy. Plast Reconstr Surg. 2009;124: 144e–55e.
14. Brown SH, Noble BC, Yang LJ, Nelson VS. Deficits in elbow position sense in neonatal brachial plexus palsy. Pediatr Neurol. 2013;49:324–8.
15. Brown SH, Wernimont CW, Phillips L, Kern KL, Nelson VS, Yang LJ. Hand sensorimotor function in older children with neonatal brachial plexus palsy. Pediatr Neurol. 2016;56:42–7.
16. Chang KW, Justice D, Chung KC, Yang LJ. A systematic review of evaluation methods for neonatal brachial plexus palsy: a review. J Neurosurg Pediatr. 2013;12(4):395–405.
17. Chang KW, Yang LJ, Driver L, Nelson VS. High prevalence of early language delay exists among toddlers with neonatal brachial plexus palsy. Pediatr Neurol. 2014;51:384–9.
18. Colbert SH, Mackinnon S. Posterior approach for double nerve transfer for restoration of shoulder function in upper brachial plexus palsy. Hand (N Y). 2006;1:71–7.

19. Dodakundi C, Doi K, Hattori Y, Sakamoto S, Fujihara Y, Takagi T, et al. Outcome of surgical reconstruction after traumatic total brachial plexus palsy. J Bone Joint Surg Am. 2013;95:1505–12.
20. Doi K, Kuwata N, Muramatsu K, Hottori Y, Kawai S. Double muscle transfer for upper extremity reconstruction following complete avulsion of the brachial plexus. Hand Clin. 1999;15:757–67.
21. Doi K, Shigetomi M, Kaneko K, Soo-Heong T, Hiura Y, Hattori Y, et al. Significance of elbow extension in reconstruction of prehension with reinnervated free-muscle transfer following complete brachial plexus avulsion. Plast Reconstr Surg. 1997;100:364–72; discussion 373–64.
22. Dy CJ, Garg R, Lee SK, Tow P, Mancuso CA, Wolfe SW. A systematic review of outcomes reporting for brachial plexus reconstruction. J Hand Surg Am. 2015;40:308–13.
23. Estrella EP, Favila Jr AS. Nerve transfers for shoulder function for traumatic brachial plexus injuries. J Reconstr Microsurg. 2014;30:59–64.
24. Flores LP. Results of surgical techniques for re-innervation of the triceps as additional procedures for patients with upper root injuries. J Hand Surg Eur Vol. 2013;38:248–56.
25. Flores LP. Triceps brachii reinnervation in primary reconstruction of the adult brachial plexus: experience in 25 cases. Acta Neurochir. 2011;153:1999–2007.
26. Garg R, Merrell GA, Hillstrom HJ, Wolfe SW. Comparison of nerve transfers and nerve grafting for traumatic upper plexus palsy: a systematic review and analysis. J Bone Joint Surg Am. 2011;93:819–29.
27. Gilbert A, Brockman R, Carlioz H. Surgical treatment of brachial plexus birth palsy. Clin Orthop Relat Res. 1991;264:39–47.
28. Gilbert A, Pivato G, Kheiralla T. Long-term results of primary repair of brachial plexus lesions in children. Microsurgery. 2006;26:334–42.
29. Gilbert A, Tassin JL. Surgical repair of the brachial plexus in obstetric paralysis. Chirurgie. 1984;110:70–5.
30. Goubier JN, Teboul F, Khalifa H. Reanimation of elbow extension with intercostal nerves transfers in total brachial plexus palsies. Microsurgery. 2011;31:7–11.
31. Hill EL. Non-specific nature of specific language impairment: a review of the literature with regard to concomitant motor impairments. Int J Lang Commun Disord. 2001;36:149–71.
32. Kennedy R. suture of the brachial plexus in birth paralysis of the upper extremity. Br Med J. 1903;1:298–301.
33. Konig RW, Antoniadis G, Borm W, Richter HP, Kretschmer T. Role of intraoperative neurophysiology in primary surgery for obstetrical brachial plexus palsy (OBPP). Childs Nerv Syst. 2006;22:710–4.
34. Little KJ, Zlotolow DA, Soldado F, Cornwall R, Kozin SH. Early functional recovery of elbow flexion and supination following median and/or ulnar nerve fascicle transfer in upper neonatal brachial plexus palsy. J Bone Joint Surg Am. 2014;96:215–21.
35. Malessy MJ, de Ruiter GC, de Boer KS, Thomeer RT. Evaluation of suprascapular nerve neurotization after nerve graft or transfer in the treatment of brachial plexus traction lesions. J Neurosurg. 2004;101:377–89.
36. Malessy MJ, Pondaag W, van Dijk JG. Electromyography, nerve action potential, and compound motor action potentials in obstetric brachial plexus lesions: validation in the absence of a "gold standard". Neurosurgery. 2009;65:A153–9.
37. Malungpaishrope K, Leechavengvongs S, Witoonchart K, Uerpairojkit C, Boonyalapa A, Janesaksrisakul D. Simultaneous intercostal nerve transfers to deltoid and triceps muscle through the posterior approach. J Hand Surg Am. 2012;37:677–82.
38. McRae MC, Borschel GH. Transfer of triceps motor branches of the radial nerve to the axillary nerve with or without other nerve transfers provides antigravity shoulder abduction in pediatric brachial plexus injury. Hand (N Y). 2012;7:186–90.
39. Miyamoto H, Leechavengvongs S, Atik T, Facca S, Liverneaux P. Nerve transfer to the deltoid muscle using the nerve to the long head of the triceps with the da Vinci robot: six cases. J Reconstr Microsurg. 2014;30:375–80.
40. Pet MA, Ray WZ, Yee A, Mackinnon SE. Nerve transfer to the triceps after brachial plexus injury: report of four cases. J Hand Surg Am. 2011;36:398–405.

41. Pondaag W, van der Veken LP, van Someren PJ, van Dijk JG, Malessy MJ. Intraoperative nerve action and compound motor action potential recordings in patients with obstetric brachial plexus lesions. J Neurosurg. 2008;109:946–54.
42. Ray WZ, Yarbrough CK, Yee A, Mackinnon SE. Clinical outcomes following brachialis to anterior interosseous nerve transfers. J Neurosurg. 2012;117:604–9.
43. Robert EG, Happel LT, Kline DG. Intraoperative nerve action potential recordings: technical considerations, problems, and pitfalls. Neurosurgery. 2009;65:A97–104.
44. Rui J, Zhao X, Zhu Y, Gu Y, Lao J. Posterior approach for accessory-suprascapular nerve transfer: an electrophysiological outcomes study. J Hand Surg Eur Vol. 2013;38:242–7.
45. Sarac C, Duijnisveld BJ, van der Weide A, Schoones JW, Malessy MJ, Nelissen RG, et al. Outcome measures used in clinical studies on neonatal brachial plexus palsy: a systematic literature review using the International Classification of Functioning, Disability and Health. J Pediatr Rehabil Med. 2015;8:167–85. ; quiz 185–166,
46. Satbhai NG, Doi K, Hattori Y, Sakamoto S. Functional outcome and quality of life after traumatic total brachial plexus injury treated by nerve transfer or single/double free muscle transfers: a comparative study. Bone Joint J. 2016;98-B:209–17.
47. Seruya M, Shen SH, Fuzzard S, Coombs CJ, McCombe DB, Johnstone BR. Spinal accessory nerve transfer outperforms cervical root grafting for suprascapular nerve reconstruction in neonatal brachial plexus palsy. Plast Reconstr Surg. 2015;135:1431–8.
48. Somashekar D, Yang LJ, Ibrahim M, Parmar HA. High-resolution MRI evaluation of neonatal brachial plexus palsy: a promising alternative to traditional CT myelography. AJNR Am J Neuroradiol. 2014;35:1209–13.
49. Squitieri L, Larson BP, Chang KW, Yang LJ, Chung KC. Understanding quality of life and patient expectations among adolescents with neonatal brachial plexus palsy: a qualitative and quantitative pilot study. J Hand Surg Am. 2013;38:2387–97. e2382
50. Strombeck C, Krumlinde-Sundholm L, Remahl S, Sejersen T. Long-term follow-up of children with obstetric brachial plexus palsy I: functional aspects. Dev Med Child Neurol. 2007;49:198–203.
51. Terzis JK, Kokkalis ZT. Restoration of elbow extension after primary reconstruction in obstetric brachial plexus palsy. J Pediatr Orthop. 2010;30:161–8.
52. Tse R, Kozin SH, Malessy MJ, Clarke HM. International Federation of societies for surgery of the hand committee report: the role of nerve transfers in the treatment of neonatal brachial plexus palsy. J Hand Surg Am. 2015;40:1246–59.
53. Tse R, Marcus JR, Curtis CG, Dupuis A, Clarke HM. Suprascapular nerve reconstruction in obstetrical brachial plexus palsy: spinal accessory nerve transfer versus C5 root grafting. Plast Reconstr Surg. 2011;127:2391–6.
54. Yang LJ, Anand P, Birch R. Limb preference in children with obstetric brachial plexus palsy. Pediatr Neurol. 2005;33:46–9.
55. Yang LJ, Chang KW, Chung KC. A systematic review of nerve transfer and nerve repair for the treatment of adult upper brachial plexus injury. Neurosurgery. 2012;71:417–29; discussion 429.

Alternative Strategies for Nerve Reconstruction

7

F. Siemers and K.S. Houschyar

Contents

F. Siemers (✉) • K.S. Houschyar
Plastic and Reconstructive Surgery, BG Clinic Bergmannstrost Halle,
Merseburger Str. 165, 06112 Halle, Germany
e-mail: frank.siemers@bergmannstrost.de; Khosrow-Houschyar@gmx.de

© Springer International Publishing AG 2017
K. Haastert-Talini et al. (eds.), *Modern Concepts of Peripheral Nerve Repair*,
DOI 10.1007/978-3-319-52319-4_7

7.1 Introduction

A peripheral nerve injury results whenever a nerve is crushed, compressed, or cut and in consequence the proper communication between the peripheral and central nervous system is lost [1]. Peripheral nerves possess the capacity of self-regeneration after traumatic injury. The quality of functional regeneration depends on a number of factors including location of the injury, size, and the age of the individual [2]. The classification of injury type is useful to understand the likelihood of complete recovery and the prognosis. The longitudinal nature of crushing injuries and different levels of nerve injury can be seen at various locations along the nerve. This is the most challenging nerve injury for the surgeon as some fascicles will need to be protected and not "downgraded," whereas others will require surgical reconstruction [3] (Table 7.1). Microsurgical reconstruction is required for reconnecting nerve ends, and if substance loss occurs, the two stumps must be bridged. Autologous nerve grafts have been the most widely used strategy for bridging nerve gaps; nonetheless, this technique has disadvantages (Table 7.2) [4]. During the last years, significant developments in materials sciences have

Table 7.1 Neurosensory recovery based upon Sunderland classification

Neurosensory recovery based upon Sunderland classification			
Sunderland	Recovery pattern	Rate of recovery	Need for surgery
I	Complete	Fast (days–weeks)	None
II	Complete	Slow (weeks)	None
III	Variable	Slow (weeks–months)	Maybe
IV	Poor	Little/None	Yes
V	None	None	Yes

Table 7.2 Advantages and disadvantages of nerve grafts for the repair of peripheral nerve injury

Nerve transplant	Advantages	Disadvantages
Nerve autograft	Provides a suitable environment for nerve regeneration	Limited amount of tissue
	No risk of immunological rejection	Limited number of grafts
	Simple and safe to obtain	Donor site morbidity and potential loss of function
	Easy to suture to the injured tissue	
Nerve allograft	Unlimited source tissue	Lack of appropriate animal donor tissue
	No donor site trauma for the recipient	Uncertain histocompatibility
		Ethical and legal concerns
Tissue-engineered material	Fabricated from polymers or biomacromolecules	Degradable biomaterials are expensive
	Unlimited source materials	Antigenicity
	Easy to produce	Exhibit poor tenacity, making them difficult to suture to the injured nerve
	No donor site trauma	

represented lively research in the area of alternative (nonnervous) conduits. There was an increasing availability of a number of new innovative manufacturing procedures and biomaterials [4]. Translation to the patient of artificial synthetic nerve grafts is still limited in spite of the large body of preclinical research. Today, the most popular approach is still biological tubulization with nonnervous autologous tissues, creating a scaffold that can bridge a nerve gap [4]. In fact, this approach avoids complications due to any possible graft-versus-host reaction and is less expensive.

7.2 Peripheral Nerve Grafts

Autografts In patients with larger nerve gaps where the injury must be bridged, use of an autograft remains the most reliable repair technique [3, 5]. Nerve autografts have been studied extensively, and their superiority over epineurial suturing under tension has been reported [6]. By using nerve autografts, the surgeons prepare a structural guidance of the natural material for axonal progression from the proximal to the distal nerve stumps. Donor sites for autograft nerve tissue are represented by functionally less important nerves like superficial cutaneous nerve, posterior interosseous nerve, sural nerves, or medial and lateral antebrachial cutaneous nerves [6, 7].

The three major types of autografts are trunk grafts, cable grafts, and vascularized nerve grafts [3]. Trunk grafts are mixed motor and sensory grafts. Trunk grafts have poor functional results due to their instability and large diameters which inhibits its ability to properly revascularize the center of the graft. Cable grafts are several sections of small nerve grafts aligned in parallel to connect fascicular groups. Vascularized nerve grafts have the advantage that there is no period of ischemia compared to nonvascularized grafts and the necessity for revascularization is avoided [3]. Sensory donor nerves are most often used, with the sural nerve being the most commonly harvested (Fig. 7.1a, b). The choice of autograft is dependent on several factors, that is, the size of the nerve gap, location of proposed nerve repair, and associated donor-site morbidity [8]. Use of autografts is currently restricted to critical nerve gaps of nearly 5 cm length [6].

The main limitations in the use of nerve autografts are considered to be mismatch of donor nerve size and fascicular inconsistency between the autograft and the distal/proximal stumps of the recipient nerve. Because a mismatch in axonal size and alignment further limits the regeneration capacity of the autografts, the type of nerve autografts chosen, like motor nerves, sensory nerves, or mixed nerves, is also decisive for a successful outcome [6]. A prolonged surgical time together with the potential risk of infection and formation of painful neuroma represents other important drawbacks of nerve autografting [6]. Altogether, the recovery time for the patient can be prolonged, owing to the need for a second surgery. The limitations of autografts forced researchers to develop alternative manufacturing approaches for novel nerve conduits for peripheral nerve repair [6].

Allografts Nerve allografts have a history that exceeds that of autografts. In 1885, Eduard Albert reported the use of a nerve allograft from an amputated limb to bridge a 3 cm median nerve gap arising from resection of a sarcoma [9]. The use of donor-related

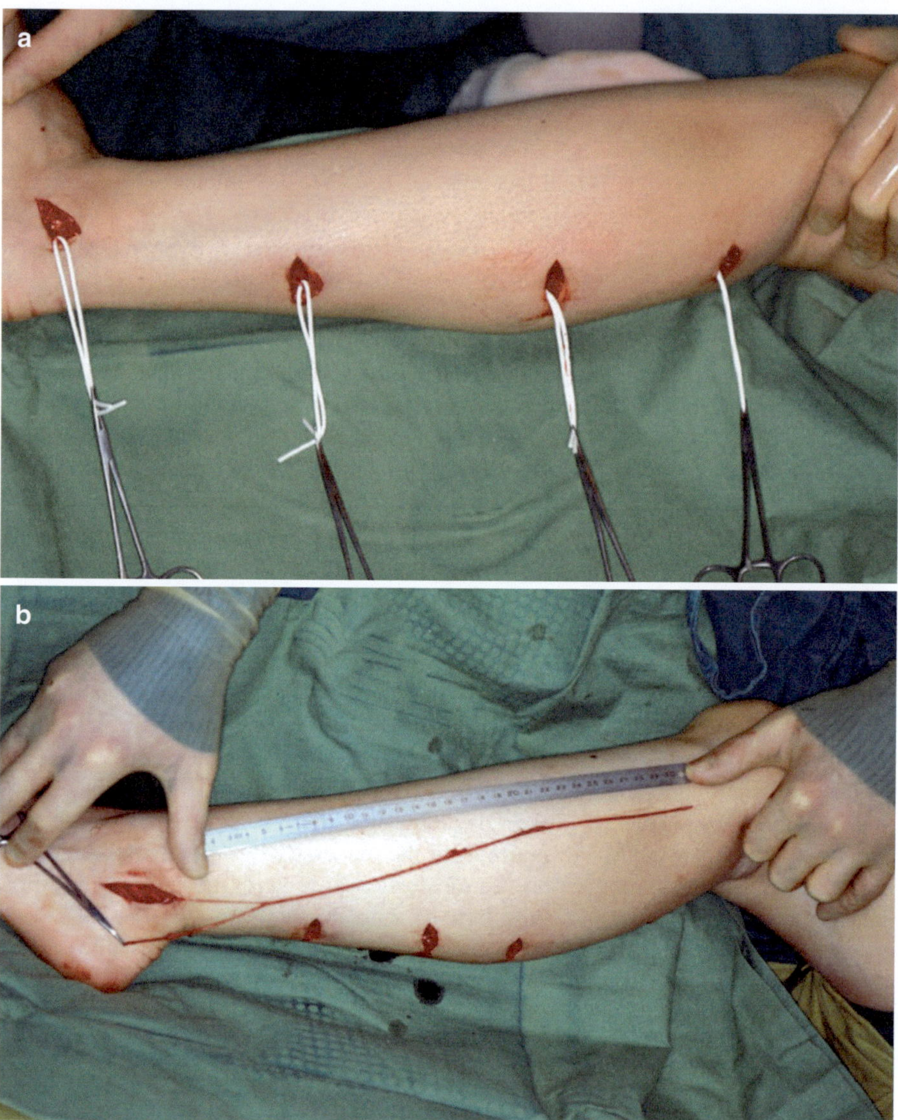

Fig. 7.1 (**a, b**) Sensory donor nerves are most often used, with the sural nerve being the most commonly harvested. (**a, b**) A minimally invasive technique of sural nerve harvesting is begun through a small incision at the level of the lateral malleolus, thereby identifying the nerve and inserting the nerve-harvesting device. An additional small incision, if needed, is placed at the junction of the middle and distal thirds of the lower leg, a landmark at which an anastomosis occurs between the medial and lateral sural cutaneous nerves

or cadaveric nerve allografts is reserved for devastating or segmental nerve injuries. Like all tissue allotransplantation, nerve allografts require systemic immunosuppression; the associated morbidity of immunomodulatory therapy limits the widespread application of nerve allografting. Several techniques (e.g., irradiation, cold preservation, lyophilization) to reduce nerve allograft antigenicity have been published [10].

Nerve allografts have demonstrated clinical utility in repairing extensive peripheral nerve injuries where there is a paucity of donor nerve material. Allografts used in peripheral nerve injuries are commercially processed to be cell- and protein-free [3]. This allows the nerve allograft to serve as a scaffold that is repopulated by Schwann cells and host axons over time [3]. The use of allografts presents limitations including especially risk of cross contamination, immune rejection, secondary infection, and limited supply [6]. Therefore, the use of allografts requires systemic immunosuppressive therapy, but long-term immune suppression is not a desirable treatment due to increased risk of infection and decrease of healing rate, and it occasionally results in tumor formation and other systemic effects [6]. In order to overcome some of these limitations, nerve allografts can be processed by repeated irradiation, freeze–thaw cycles, and decellularization with detergents [11].

7.3 Allogeneic Decellularized Nerve Transplantation

Since 2007, decellularized nerve grafts are in clinical use. Since 2013, this alternative is also available in German-speaking countries; however, only a few clinics in Germany gathered experience in this field (Fig. 7.2).

The allogeneic transplants, which are generated from human donor nerve, combine many advantages due to its macrostructure and the three-dimensional microstructure. First clinical observations indicated in broken sensitive nerves good results to a defect distance of 3 cm. In the ten cases described, the 2-point discrimination (2PD) was 6 mm or better.

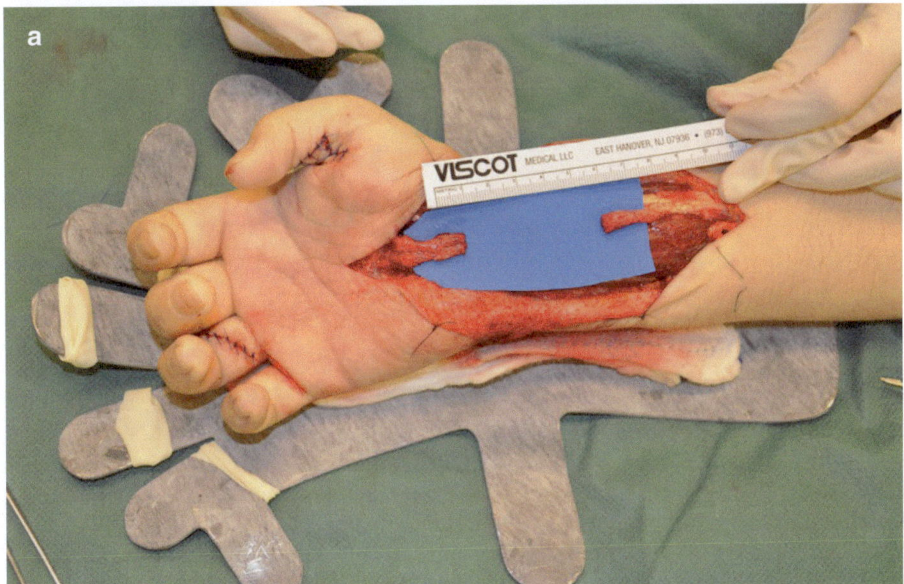

Fig. 7.2 Allogeneic decellularized nerve transplantation. (**a**) Surgical preparation of the median nerve. (**b**) Bridging the nerve gap with a decellularized allogeneic nerve. (**c**) Suturing the decellularized qallogeneic nerve with the median nerve

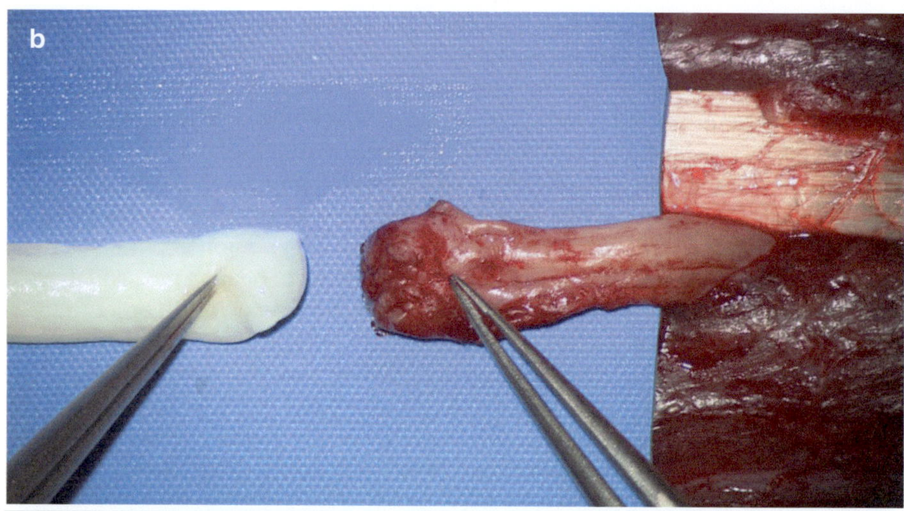

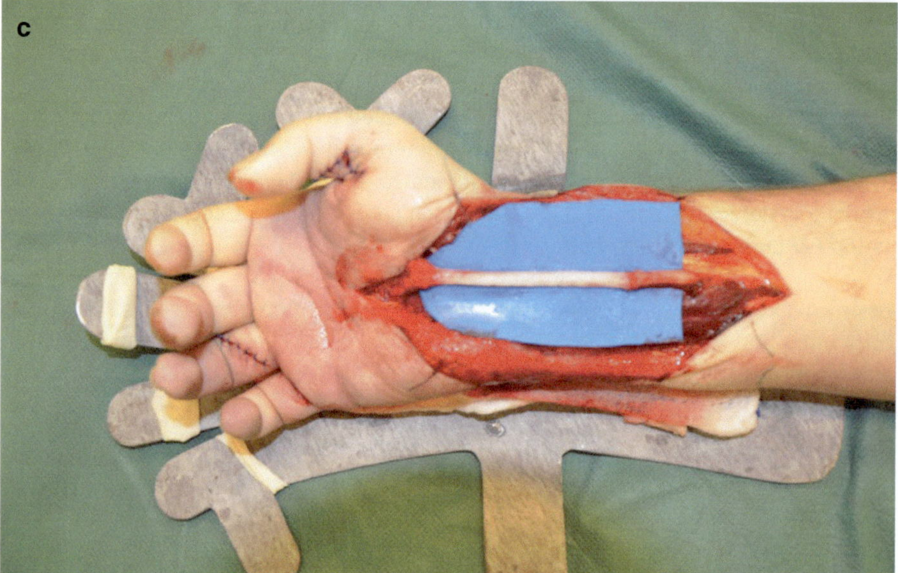

Fig. 7.2 (continued)

The largest prospective study on the use of allogeneic nerve grafts was published in 2011 by Brooks et al. under the name "RANGER study." The RANGER study registry was initiated in 2007 to study the use of processed nerve allografts (AxoGen® nerve allograft (AxoGen Inc., Alachua, FL)) in contemporary clinical practice [12]. Twelve sites with 25 surgeons contributed data from 132 individual nerve injuries. Data was analyzed to determine the safety and efficacy of the nerve allograft. Sufficient data for efficacy analysis were reported in 76 injuries (49 sensory, 18 mixed, and 9 motor nerves). The mean age was 41 ± 17 (18–86) years. The mean graft length was

22 ± 11 (5–50) mm. Subgroup analysis was performed to determine the relationship to factors known to influence outcomes of nerve repair such as nerve type, gap length, patient age, time to repair, age of injury, and mechanism of injury. Meaningful recovery was reported in 87% of the repairs reporting quantitative data. No graft-related adverse experiences were reported, and a 5% revision rate was observed. Processed nerve allografts performed well and were found to be safe and effective in sensory, mixed, and motor nerve defects between 5 and 50 mm. The outcomes for safety and meaningful recovery observed in this study compare favorably to those reported in the literature for nerve autograft and are higher than those reported for nerve conduits.

Xenografts A nerve xenograft is obtained from a member of a species other than that of the recipient. A research group developed an experimental animal model to study the potential transplantation of nerve xenografts using the newer immunosuppressive agents RS-61443 and FK-506. They transplanted 2 cm sciatic nerve xenografts obtained from golden Syrian hamsters into a 0.5 cm gap in the sciatic nerve of Lewis rats. The functional recovery in the test animals was found to be not as good as those in the control autografts [13].

Another research team used acellular nerve xenografts and seeded them with bone marrow stromal cells [14]. When the allograft and the xenograft were compared with electrophysiological studies, it was observed that the xenografts were as effective as the allografts in regenerating the nerves. Allograft and xenografts have certain disadvantages such as disease transmission and immunogenicity.

7.4 Nerve Conduits

7.4.1 Biological Nerve Conduits

The use of a conduit as a vehicle for moderation and modulation of the cellular and molecular ambience for nerve regeneration has been widely investigated [15].

A combination of physical, biological, and chemical factors has made the study of nerve tubes a complex process, rising tremendous interest in the fields of medicine. The ideal tubular material has not yet been established. Several materials, either of biologic origin or synthetically fabricated, have been applied for these purposes. The ideal conduit would be made of a low-cost, biologically inert material that is biocompatible; flexible; thin; transparent; inhibitor of inflammatory processes such fibrosis, neuromas, gliomas, swelling, ischemia, and adhesions; and facilitator of the processes that contribute to regeneration, accumulating factors that promote nerve growth [15]. Biological conduits such as autologous veins, arteries, muscle, and heterogeneous collagen tubes denatured skeletal muscle or muscle basal lamina, veins, and polyglycolic acid (PGA)–collagen tubes [16]. Biomaterials such as artery, vein, and muscle have been widely used to repair relatively short nerve defects. These materials can provide support for the nerve in the short term and degrade to innocuous products after complete nerve regeneration.

Table 7.3 Design criteria for nerve guidance conduits

Ideal properties	Description
Biocompatibility	Material should not harm the surrounding tissues
Protein modification/release	Laminin/fibronectin coating for increased cellular adhesion; controlled/sustained growth factor release
Degradation/porosity	Degradation rate should complement nerve regeneration rate; conduit should allow nutrient diffusion and limit scar tissue infiltration
Anisotropy	An internal scaffold or film should provide directional guidance
Physical fit	Conduit should have a large enough internal diameter to not "squeeze" the regenerating nerve; wall thickness limited
Electrically conducting	Capable of propagating electrical signals
Support cells	Schwann cells/stem cells capable of delivering neurotrophic factors to the site of regeneration

Table 7.3 summarizes the types and the performance of a variety of conduit materials.

In the 1980s, preclinical research by Glasby and colleagues demonstrated that autografts of skeletal muscle which had been deeply frozen in liquid nitrogen and subsequently thawed can provide a valuable matrix for the regenerating nerve, when oriented coaxially with respect to the nerve tissue. In eight patients, this type of grafts was used to repair injured digital nerves. Assessment from 3 to 11 months after operation showed recovery to MRC (Medical Research Council) sensory category S3+ in all patients [17]. Lundborg reported about different methods of frozen muscle grafts and other conduits bridging nerve gaps [18].

There are several advantages, however, in using vein conduits for nerve reconstruction [19]. The tissue composition of veins is similar to that of nerve tissue. Furthermore, muscle–vein-combined graft conduits have been broadly devised and effectively employed for repair of segmental nerve injuries [20]. Manoli et al. conducted a retrospective clinical trial in order to compare regeneration results after digital nerve reconstruction with muscle-in-vein conduits, nerve autografts, or direct suture [21]. In a total of 46 patients with 53 digital nerve injuries with a segmental nerve injury ranging between 1 and 6 cm, no statistically significant differences between all three groups could be found. The authors also emphasized that after harvesting a nerve graft, reduction of sensibility at the donor site occurred in 10 of 14 cases but only in one case after harvesting a muscle-in-vein conduit.

7.4.2 Synthetic Nerve Conduits

They include nondegradable and degradable nerve conduits (Fig. 7.3). Synthetic polymers, though often less biocompatible relative to biopolymers, offer opportunities for tailored degradation and control of mechanical strength, porosity, and microstructure properties [22].

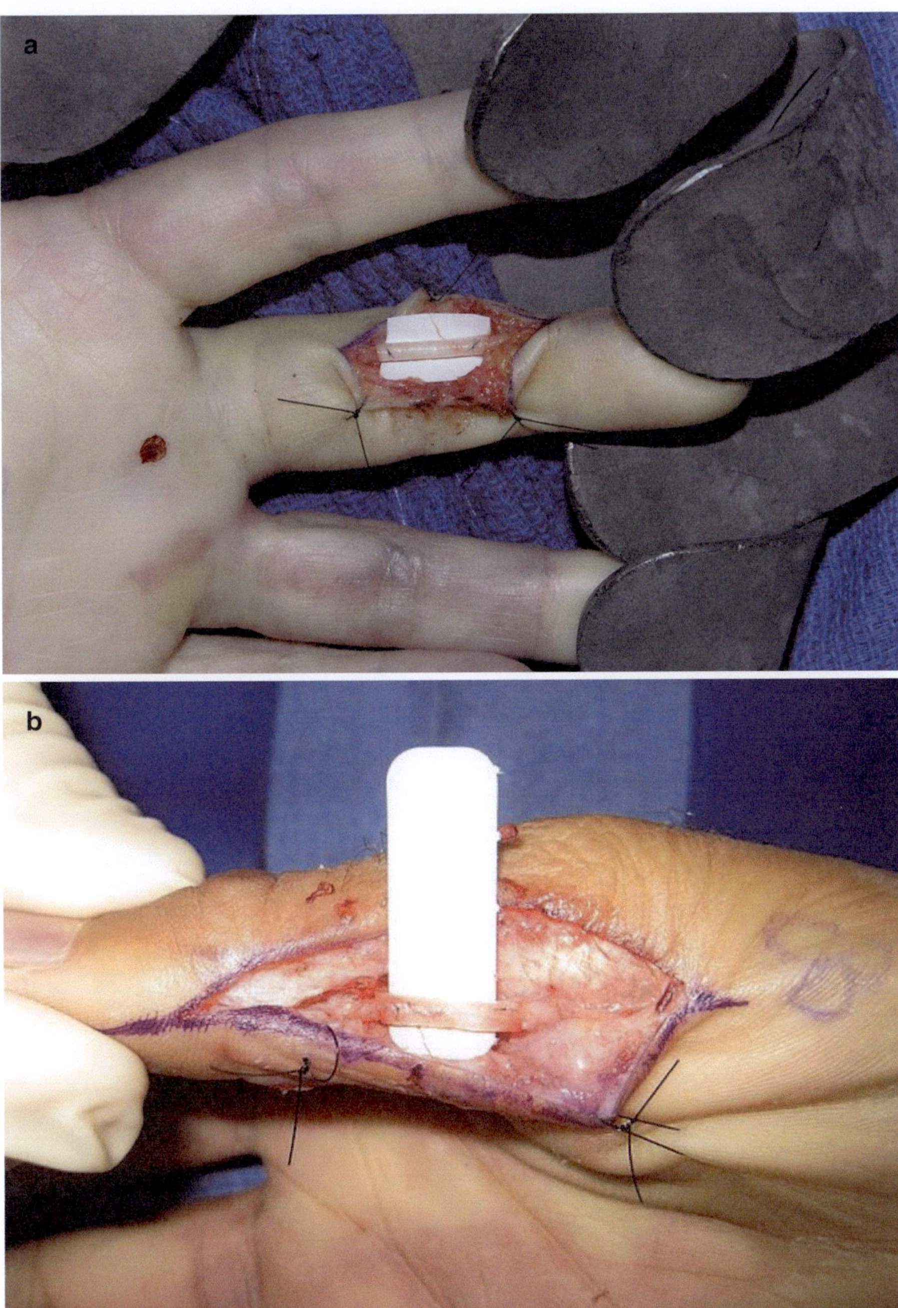

Fig. 7.3 Examples for nerve conduit designs. (**a**) Acellular nerve repair material and its application for the repair of 2 cm ulnar nerve defect in the fourth finger. (**b**) Acellular nerve repair material and its application for the repair of 2 cm radial nerve defect in the thumb

In the meantime, the material choice for nerve conduits shifted toward the use of more biocompatible synthetic polymers. Biodegradable polyesters, such as polyglycolic acid (PGA), polylactic acid (PLA), poly(ε-caprolactone) (PCL), poly(lactic acid-co-glycolic acid) (PLGA), polyurethanes (PUs), and nonbiodegradable polymers such as methacrylate-based hydrogels, silicone, polystyrene, and polytetrafluoroethylene), were used as nerve conduit materials and intensively studied in preclinical models [6].

7.4.3 Technique of Tubulization (Fig. 7.4)

Surgery on the peripheral nerve requires microsurgical techniques. Following debridement and neurolysis if applicable, the nerve stumps are located with one or two u-sutures of 8/0 to 10/0 nylon and inserted into the moistened tube with an overlap of 2–3 mm (Fig. 7.4a, c).

In addition to the defect length, the nerve diameter must be detected in the preparation. Of this, the choice of the size of the nerve conduit depends. All operational steps have to be carried out using a magnifying optics and microsurgical instruments and sutures. A microsurgical expertise is a prerequisite for a successful placement of a nerve tube [23].

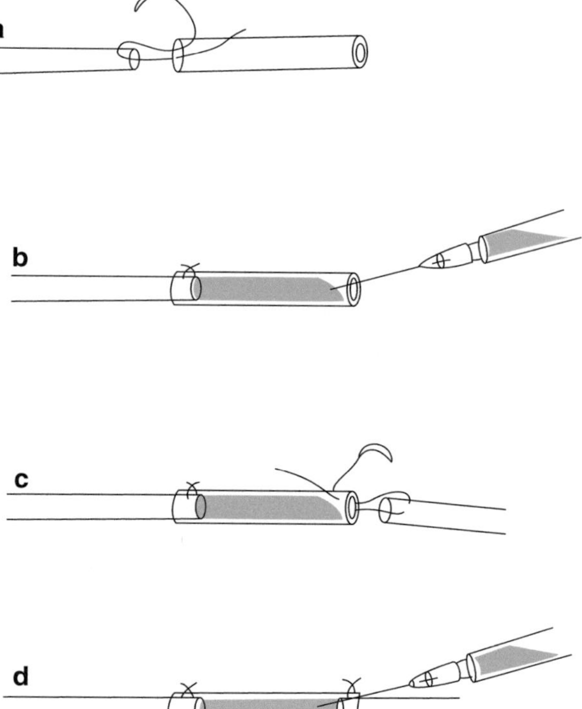

Fig. 7.4 Schematic representation of the tubulization technique. (**a–c**) Nerve stumps are located with one or two u-sutures of nylon and inserted into the moistened tube with an overlap of 2–3 mm. (**b, d**) After finishing each coaptation, the lumen has to be rinsed with normal saline or electrolyte solution using a small cannula to remove any remaining blood clots

Tubulization has to be performed after release of the tourniquet to prevent bleeding into the conduit. After finishing each coaptation, the lumen has to be rinsed with normal saline or electrolyte solution using a small cannula to remove any remaining blood clots (Fig. 7.4b, d). Immobilization of the adjacent joints is advisable for at least 10 days. Massaging the scar should be avoided due to the risk of dislocation of the tube in the first weeks following the operation. Tubulization seems equally appropriate for primary and secondary nerve reconstructions as well as for reconstruction after neuroma resection [24].

7.5 Polyesters for Nerve Conduit Fabrication

Most of current resorbable synthetic polymer membranes on the market are based on aliphatic polyesters. PGA, PLA, PLLA, PLGA, and PCL are polyesters most commonly used in the fabrication of nerve conduits.

7.5.1 PGA

The PGA conduit, also known as the GEM Neurotube, has been the most extensively studied synthetic biodegradable conduit both experimentally and clinically [20]. First descriptions go back to Mackinnon and Dellon who published in 1990 a report of 15 digital nerve lesions being reconstructed with hollow polyglactin (PGA) conduits [25]. It is a porous synthetic aliphatic polyester made of polyglycolic acid, which exhibits a high tensile modulus with very low solubility in organic solvents [26]. In an earlier study, PGA-based crimped tube device (Neurotube®; Synovis Micro Companies Alliance, Birmingham, AL, USA) was described for the repair of peripheral nerve injuries [6]. In a more recent experimental study, bone marrow-derived stem cells (BMSCs) were combined with PGA tube (PGAt) (Neurotube®) in autografted rat facial nerves [27]. After cutting of the mandibular branch of the rat facial nerve, surgical repair consisted of autologous graft in a PGA filled with basement membrane matrix with undifferentiated bone marrow-derived stem cells (BMSCs) or Schwann-like cells that had been differentiated from BMSCs. After 6 weeks of surgery, animals from either cell-containing group had compound muscle action potential amplitudes significantly higher than the control groups. PGA is also often combined with natural polymers such as collagen [28]. Weber et al. reported the results of the first randomized, prospective, multicenter evaluation comparing autografts and PGA conduits for the repair of digital nerve gaps [29]. PGA tubes produced good to excellent functional sensation in 100% of patients with nerve gaps <4 mm, 83% of patients with nerve gaps 5–7 mm, and 71% of patients with nerve gaps >8 mm.

7.5.2 PLA

PLA (polylactic acid) is one of the most common and important polymers because of its suitable mechanical properties and biocompatibility [30]. Biocompatible

PLA can be derived from lactic acid obtained from corn, sugar beet, or wheat. PLA has been used commercially as membranes, such as Resolut Adapt®, Vicryl®, Epi-Guide®, and Vivosorb®, and each of these membranes may have its own properties. PLA was used as a nerve conduit material in a number of studies [31]. In one study, a multilayer PLA nerve conduit was fabricated by microbraiding to obtain adequate mechanical strength at the injury site [32]. In the experimental applications on rats, successful regeneration through a 10 mm gap was observed at 8 weeks after operation. In another study, a PLA nerve conduit was made by immersion precipitation to bridge a 20 mm long gap in an animal nerve transection model [33]. The researchers reported that the functional recovery after 18 months was about 80%, based on electrophysiology and behavior analysis. PLA conduits grafted with FGF1 (fibroblast growth factor 1) and chitosan–nano-Au (gold) after plasma activation showed the greatest regeneration capacity and functional recovery when they were tested for their ability to bridge a 15 mm critical gap defect in a rat sciatic nerve injury model.

7.5.3   PLLA

PLLA is a highly crystalline and stereoregular form of PLA. Researchers reported that PLLA nerve conduits modified with laminin-derived AG73 peptides and a PEG (polyethylene glycol) containing an outer layer are effective for preventing the adhesion of surrounding tissue [34]. In one experimental study, a PLLA nerve conduit was fabricated by extrusion and was used in a 10 mm sciatic nerve defect model in rats [35]. As a result at 16 weeks, the nerve fiber density in the distal sciatic nerve repaired with the PLLA conduits was similar to that repaired with control isografts. The research group found also an increased axon number and nerve fiber density in the PLLA mid-conduit compared with control isograft at 16 weeks.

7.5.4   PLGA

PLGA (poly(lactic-co-glycolic acid)) has been the most frequently used biodegradable polymer in tissue engineering for fabricating porous foams for biomedical applications. PLGA is a co-polyester that has been evaluated extensively as a nerve guide material due to its ease of fabrication, approval by the FDA, and low inflammatory response it created [7]. In an earlier experimental study, PLGA conduits with longitudinally aligned channels were produced by using a combined thermally induced phase transition technique and injection molding [36]. Macropores were organized into bundles of channels up to 20 μm wide in the PLGA matrix, which then was used as a nerve conduit.

7.5.5 PCL

PCL (poly (ε-caprolactone)), another polyester, has high solubility in organic solvents and low melting temperature (55–60 °C) and glass transition temperatures (−60 °C) [26]. Oliveira et al. fabricated PCL conduits for regeneration of transected mouse median nerves and investigated the effect of transplanted MSCs (mesenchymal stem cells) on nerve regeneration by seeding MSCs on the PCL nerve conduits before grafting [37]. The animals treated with MSCs had a significantly larger number of regenerated unmyelinated and myelinated nerve fibers and blood vessels compared to the control group, indicating the possibility of improving regeneration and function of median nerve after a traumatic lesion.

7.5.6 Poly(D,L-lactide-co-ε-caprolactone)

Poly(D,L-lactide-co-ε-caprolactone) is a copolymer of caprolactone monomers and lactic acid. Cylindrical poly(D,L-lactide-co-ε-caprolactone) 80/20 copolymer nerve conduits were fabricated by using an ink-jet system in an experimental study [6]. Radulescu et al. found that hNGF-EcR-293 cells could be genetically modified to deliver NGF (nerve growth factor) in vitro and in vivo to support the local neuroreparative factor delivery via a tightly controlled system [38]. They demonstrated that these cells could attach and survive for more than 8 weeks when cultured on the 80/20 PLA-PCL copolymer but failed to attach and died on 25/75 and 40/60 PLA–PCL copolymer used. In another study by Chiriac et al., the Neurolac™ nerve conduit was tested in a clinical setting on 28 nerve lesions on various sites: arm, elbow, forearm, wrist, palm, and fingers with an average defect length of 11 mm [39]. After an average of 21.9 months of follow-up (3–45 months), subjective criteria (cold intolerance and pain) and objective criteria (strength) were compared with the contralateral side. Grip strength averaged 64.62% of the contralateral side. The researchers observed eight complications, the most serious being two fistulizations of the Neurolac™ device close to a joint and one neuroma formation. All in all, it was concluded that the use of Neurolac™ in repairing hand nerve defects cannot be considered very effective.

7.6 Proteins with Synthetic Biomaterials

Proteins such as collagen are natural polymers. The blends of natural polymers with synthetic polymers are considered as hybrid structures. Schmauss et al. analyzed the nerve regeneration of their patients after reconstruction with collagen nerve conduits terminated after 12 months [40]. The researchers examined 20 reconstructed nerves in 16 patients with a mean follow-up of 58.1 months (range, 29.3–93.3 months).

They found an improved sensibility at current follow-up compared with the 12-month follow-up in 13 cases. Three cases had the same values, whereas four cases had worsened sensibility. Improvement of sensibility was associated with a significantly shorter nerve gap length with significantly better results if the gap length was <10 mm. In another prospective cohort study, the clinical use of artificial nerve conduits for digital nerve repair was presented [41]. The researchers presented their clinical experiences based on a review of the outcome and techniques in the current literature. Fifteen digital nerve lesions in 14 patients have been overcome by interpositional grafting of a hollow collagen I conduit. A follow-up of 12 months could be guaranteed in 12 cases. The mean nerve gap was 12.5 ± 3.7 mm. Four out of 12 patients, assessed 12 months postoperatively, showed excellent sensibility (S4). Five patients achieved good sensibility, one poor, and two no sensibility. Lohmeyer et al. presented a prospective two-center cohort study on digital nerve reconstruction with collagen nerve conduits [42]. The data were put into the context of a comprehensive review of existing literature. Over a period of 3 years, all consecutive digital nerve lesions that could not be repaired by tensionless coaptation with a gap length of less than 26 mm were reconstructed with nerve conduits made from bovine collagen I. Sensibility was assessed 1 week and 3, 6, and 12 months postoperatively by static and moving 2-point-discrimination (2PD) and monofilament testing. Forty-nine digital nerve lesions in 40 patients met the inclusion criteria. The mean nerve gap was 12.3 ± 2.3 mm (span 5–25 mm). Forty nerve reconstructions could be included in the 12-month follow-up. Three cases, assessed 12 months postoperatively, showed excellent sensibility (static 2PD < 6 mm). Seventeen achieved good (2PD 6–10 mm), 5 fair (2PD 11–15 mm), 6 poor (2PD > 15 mm, but protective sensibility), and 9 achieved no sensibility. Monofilament test results were significantly better if gap length was shorter than 12 mm. Boeckstyns et al. demonstrated in a prospective randomized trial with 43 patients the reparation of the ulnar or the median nerve with a collagen nerve conduit or with conventional microsurgical techniques [43]. As a result, use of a collagen conduit produced recovery of motor and sensory functions that were equivalent to direct suture 24 months after repair when the nerve gap inside the tube was 6 mm or less.

7.6.1 Nondegradable Nerve Conduits: Silicone, Plastic, and Polytetrafluoroethylene Tubes

The silica gel canal was the earliest artificial conduit described in 1982 [44]. Nondegradable nerve conduits generally eliminate the need to harvest autologous nerves. But nondegradable nerve conduits always cause compression of the regenerating nerve that could negatively affect axonal regeneration, and they often cause inflammation of the surrounding tissues. Furthermore, these types of conduits require a second surgery for their removal, which could cause more injury to the patient and pain.

7.6.2 Degradable Nerve Conduits

The commonly used degradable materials include chitosan, collagen, polyglycolic acid conduit, glycolide trimethylene carbonate conduit, polylactic acid conduit, polycarbolacton conduit, natural collagen, and hydrogel conduit [45].

Researchers are enthusiastically investigating new biodegradable materials with excellent chemical and physical properties. Biodegradable collagen and chitosan collagen tubes were proved to promote the growth of axons. During the last 25 years, studies on chitosan as a biomaterial for nerve tissue engineering applications have been intensified [46].

Chitosan, a polysaccharide, which is industrially produced (hydrolyzed) from chitin, due to its high biocompatibility and stimulating influence on natural healing processes is of particular interest for use as nerve conduit. So far, conducted animal studies showed that the chitosan due to the bioactive properties supports nerve regeneration [45, 46]. The positively charged biopolymer interacts with negatively charged biomolecules and cellular components, thus promoting the restoration of the nerve continuity. Chitosan – thanks to its anti-adhesive and antibacterial properties – also ensures for a reduced formation of scar tissue and reduces the risk of infection. Clinical trial results, however, for new product chitosan nerve conduits (Reaxon® nerve guides) are still lacking. The researcher group of Haastert-Talini reported an analysis of chitosan nerve guides (CNGs) enhanced by introduction of a longitudinal chitosan film to reconstruct critical length 15 mm sciatic nerve defects in rats [47]. The investigations demonstrated that the CNGs (chitosan nerve guides) enhanced by the guiding structure of the introduced chitosan film significantly improved morphological and functional results of nerve regeneration in comparison with simple hollow CNGs. In another study, Haastert-Talini et al. showed an analysis of chitosan tubes used to reconstruct 10 mm nerve defects in rats [48]. Investigations were performed demonstrating that the chitosan tubes allowed morphological and functional nerve regeneration similar to autologous nerve grafts. Hollow biodegradable materials can be used to repair only relatively short nerve defects, and the functional recovery is still not satisfying. Neubrech et al. recently initiated a randomized double-blind controlled multicenter trial in Germany including 100 patients with traumatic sensory nerve lesions of the hand without a gap [49]. Patients will be randomized to primary microsurgical repair of the injured nerve with the additional use of a chitosan nerve tube or direct tension-free microsurgical repair of the injured nerve alone.

Furthermore, currently at three hospitals (BG hospitals) in Germany, again, a prospective randomized study has been initiated in which the results of the application of Reaxon® nerve guides will be examined. As participating clinic, we can report here as a preliminary result that our short-term investigation after 6 months demonstrates that the chitosan tubes allowed functional and morphological nerve regeneration similar to autologous nerve grafts with a nerve gaps of 6–10 mm. This suggests that similar to animal studies, chitosan nerve conduits support nerve cell adhesion and neurite outgrowth in human patients.

7.7　Conclusions and Recommendations

It can be seen that the development of nerve conduits for peripheral nerve repair is a highly sophisticated and active process. Both conduits and acellular allografts are useful tools for dealing with short nerve gaps. The convenience of either one should facilitate adequate nerve debridement and the avoidance of over tensioned repairs. Though with time and mounting experience the recommended maximum repair length of conduits seems to be decreasing while that of the acellular allograft seems to be increasing, the critical gap sizes for either tool are not known. Autograft is still the gold standard, but in the right situations, either conduits or acellular allograft can achieve equivalent or at least similar results making them excellent options for nonessential nerve repairs and something that should be at least considered for more important nerves. Though autograft donor deficits or complications are typically minimal or rare, significant problems can occur. The exact roles of both tools in the nerve repair algorithm continue to be defined.

References

1. Fex Svennigsen A, Dahlin LB. Repair of the peripheral nerve-remyelination that works. Brain Sci. 2013;3:1182–97.
2. Pfister LA, Papaloïzos M, Merkle HP, et al. Nerve conduits and growth factor delivery in peripheral nerve repair J Peripher Nerv Syst. 2007;12(2):65–82. Review.
3. Houschyar KS, Momeni A, Pyles MN, et al. The role of current techniques and concepts in peripheral nerve repair. Plast Surg Int. 2016;2016:4175293.
4. Geuna S, Tos P, Titolo P, et al. Update on nerve repair by biological tubulization. J Brachial Plex Peripher Nerve Inj. 2014;1:3.
5. Martins RS, Bastos D, Siqueira MG, et al. Traumatic injuries of peripheral nerves: a review with emphasis on surgical indication. Arq Neuropsiquiatr. 2013;71(10):811–4.
6. Arslantunali D, Dursun T, Yucel D, et al. Peripheral nerve conduits: technology update. Med Devices (Auckl). 2014;7:405–24.
7. Stang F, Stollwerck P, Prommersberger KJ, et al. Posterior interosseus nerve vs. medial cutaneous nerve of the forearm: differences in digital nerve reconstruction. Arch Orthop Trauma Surg. 2013;133(6):875–80.
8. Ray WZ, Mackinnon SE. Management of nerve gaps: autografts, allografts, nerve transfers, and end-to-side neurorrhaphy. Exp Neurol. 2010;1:77–85.
9. Albert E. Einige Operationen an Nerven. Wien Med Presse. 1885;26:1285–8.
10. Mackinnon SE, Hudson AR, Falk RE, et al. Peripheral nerve allograft: an immunological assessment of pretreatment methods. Neurosurgery. 1984;14(2):167–71.
11. Johnson PJ, Wood MD, Moore AM, et al. Tissue engineered constructs for peripheral nerve surgery. Eur Surg. 2013;45 doi:10.1007/s10353-013-0205-0.
12. Cho MS, Rinker BD, Weber RV, et al. Functional outcome following nerve repair in the upper extremity using processed nerve allograft. J Hand Surg [Am]. 2012;37(11):2340–9.
13. Hebebrand D, Zohman G, Jones NF. Nerve xenograft transplantation. Immunosuppression with FK-506 and RS-61443. J Hand Surg (Br). 1997;22(3):304–7.
14. Jia HJ, Wang Y, Tong XJ, et al. Sciatic nerve repair by acellular nerve xenografts implanted with BMSCs in rats xenograft combined with BMSCs. Synapse. 2012;66:256–69.
15. Sabongi RG, Fernandes M, Dos Santos JB, et al. Peripheral nerve regeneration with conduits: use of vein tubes. Neural Regen Res. 2015;10(4):529–33.

16. Muheremu A, Ao Q. Past, present, and future of nerve conduits in the treatment of peripheral nerve injury. Biomed Res Int. 2015;2015:237507.
17. Norris RW, Glasby MA, Gattuso JM, et al. Peripheral nerve repair in humans using muscle autografts. A new technique. J Bone Joint Surg Br. 1988;70(4):530–3.
18. Lundborg G. Neurotropism, frozen muscle grafts and other conduits. J Hand Surg (Br). 1991;16(5):473–6. Review.
19. Ahmad I, Akhtar MS. Use of vein conduit and isolated nerve graft in peripheral nerve repair: a comparative study. Plast Surg Int. 2014;2014:587968.
20. Grinsell D, Keating CP. Peripheral nerve reconstruction after injury: a review of clinical and experimental therapies. Biomed Res Int. 2014;2014:698256.
21. Manoli T, Schulz L, Stahl S, et al. Evaluation of sensory recovery after reconstruction of digital nerves of the hand using muscle-in-vein conduits in comparison to nerve suture or nerve autografting. Microsurgery. 2014;34:608–15.
22. Nectow AR, Marra KG, Kaplan DL. Biomaterials for the development of peripheral nerve guidance conduits. Tissue Eng Part B Rev. 2012;18(1):40–50.
23. Siemers F. Nervenröhrchen in der Nervenchirurgie. Trauma Berufskrankh. 2016;18(suppl 3): 260–3.
24. Siemers F. Alternativen zur autologen Nerventransplantation. Obere Extremität. 2015;10(3): 162–7.
25. Mackinnon SE, Dellon AL. Clinical nerve reconstruction with a bioabsorbable polyglycolic acid tube. Plast Reconstr Surg. 1990;85(3):419–24.
26. Madhavan Nampoothiri K, Nair NR, John RP. An overview of the recent developments in polylactide (PLA) research. Bioresour Technol. 2010;101(22):8493–501.
27. Costa HJ, Ferreira Bento R, Salomone R. Mesenchymal bone marrow stem cells within polyglycolic acid tube observed in vivo after six weeks enhance facial nerve regeneration. Brain Res. 2013;1510:10–21.
28. Ulery BD, Nair LS, Laurencin CT. Biomedical applications of biodegradable polymers. J Polym Sci B Polym Phys. 2011;49(12):832–64.
29. Weber RA, Breidenbach WC, Brown RE. A randomized prospective study of polyglycolic acid conduits for digital nerve reconstruction in humans. Plast Reconstr Surg. 2000;106(5):1036–45.
30. Asghari F, Samiei M, Adibkia K, et al. Biodegradable and biocompatible polymers for tissue engineering application: a review. Artif Cells Nanomed Biotechnol. 2016.
31. Mobasseri A, Faroni A, Minogue BM, et al. Polymer scaffolds with preferential parallel grooves enhance nerve regeneration. Tissue Eng Part A. 2015;21(5–6):1152–62.
32. Lu MC, Huang YT, Lin JH, et al. Evaluation of a multi-layer micro-braided polylactic acid fiber-reinforced conduit for peripheral nerve regeneration. J Mater Sci Mater Med. 2009;20: 1175–80.
33. Hsu SH, Chan SH, Chiang CM, et al. Peripheral nerve regeneration using a microporous polylactic acid asymmetric conduit in a rabbit long-gap sciatic nerve transection model. Biomaterials. 2011;32:3764–75.
34. Kakinoki S, Nakayama M, Moritan T, et al. Three-layer microfibrous peripheral nerve guide conduit composed of elastin-laminin mimetic artificial protein and poly(L-lactic acid). Front Chem. 2014;2:52.
35. Evans GRD, Brandt K, Widmer MS, et al. In vivo evaluation of poly(L-lactic acid) porous conduits for peripheral nerve regeneration. Biomaterials. 1999;20:1109–15.
36. Gaudin R, Knipfer C, Henningsen A, et al. Approaches to peripheral nerve repair: generations of biomaterial conduits yielding to replacing autologous nerve grafts in craniomaxillofacial surgery. Biomed Res Int. 2016;2016:3856262. doi: 10.1155/2016/3856262. Epub 2016 Jul 31.
37. Oliveira JT, Almeida FM, Biancalana A, et al. Mesenchymal stem cells in a polycaprolactone conduit enhance median-nerve regeneration, prevent decrease of creatine phosphokinase levels in muscle, and improve functional recovery in mice. Neuroscience. 2010;170(4): 1295–303.

38. Radulescu D, Dhar S, Young CM, et al. Tissue engineering scaffolds for nerve regeneration manufactured by ink-jet technology. Mater Sci Eng C. 2007;27(3):534–9.
39. Chiriac S, Facca S, Diaconu M, et al. Experience of using the bioresorbable copolyester poly(DL-lactide-ε-caprolactone) nerve conduit guide Neurolac™ for nerve repair in peripheral nerve defects: report on a series of 28 lesions. J Hand Surg Eur. 2011;37:342–9.
40. Schmauss D, Finck T, Liodaki E, et al. Is nerve regeneration after reconstruction with collagen nerve conduits terminated after 12 months? The long-term follow-up of two prospective clinical studies. J Reconstr Microsurg. 2014;30(8):561–8.
41. Lohmeyer JA, Siemers F, Machens HG, et al. The clinical use of artificial nerve conduits for digital nerve repair: a prospective cohort study and literature review. J Reconstr Microsurg. 2009;25(1):55–61.
42. Lohmeyer JA, Kern Y, Schmauss D, et al. Prospective clinical study on digital nerve repair with collagen nerve conduits and review of literature. J Reconstr Microsurg. 2014;30(4):227–34.
43. Boeckstyns ME, Sørensen AI, Viñeta JF, et al. Collagen conduit versus microsurgical neurorrhaphy: 2-year follow-up of a prospective, blinded clinical and electrophysiological multicenter randomized, controlled trial. J Hand Surg [Am]. 2013;38(12):2405–11.
44. Lundborg G, Dahlin LB, Danielsen N, et al. Nerve regeneration in silicone chambers: influence of gap length and of distal stump components. Exp Neurol. 1982;76(2):361–75. doi:10.1016/0014-4886(82)90215-1.
45. Wangensteen KJ, Kalliainen LK. Collagen tube conduits in peripheral nerve repair: a retrospective analysis. Hand. 2010;5(3):273–7. doi:10.1007/s11552-009-9245-0.
46. Gnavi S, Barwig C, Freier T, et al. The use of chitosan-based scaffolds to enhance regeneration in the nervous system. Int Rev Neurobiol. 2013;109:1–62.
47. Meyer C, Stenberg L, Gonzalez-Perez F, et al. Chitosan-film enhanced chitosan nerve guides for long-distance regeneration of peripheral nerves. Biomaterials. 2016;76:33–51.
48. Haastert-Talini K, Geuna S, Dahlin LB, et al. Chitosan tubes of varying degrees of acetylation for bridging peripheral nerve defects. Biomaterials. 2013;34(38):9886–904.
49. Neubrech F, Heider S, Harhaus L, et al. Chitosan nerve tube for primary repair of traumatic sensory nerve lesions of the hand without a gap: study protocol for a randomized controlled trial. Trials. 2016;17:48.

Clinical Follow-Up

8

Gilda Di Masi, Gonzalo Bonilla, and Mariano Socolovsky

Contents

Analysing the results of nerve repair is very important to compare the effectiveness of different strategies and, thus, develop standardized guidelines for the management and treatment of nerve injuries.

This analysis process can be extremely complex, since many different factors influence functional recovery after peripheral nerve repair. Two of the most important variables are the time between the injury and the surgery, and the level of repair. The deleterious effect of time is widely reflected in the literature [6, 10, 22], no matter which technique is employed during nerve reconstruction. Within 3 weeks of denervation, muscle atrophy begins, and over the next 2 years, the muscle is almost totally replaced by fibrous tissue. If the period to reinnervation of the main effectors

G. Di Masi (✉) • G. Bonilla • M. Socolovsky (✉)
Peripheral Nerve and Plexus Surgery Unit, University of Buenos Aires School of Medicine, Hospital de Clínicas, Buenos Aires, Argentina
e-mail: gildadimasi@gmail.com; bonilla_gonzalo@hotmail.com; marianosocolovsky@gmail.com

© Springer International Publishing AG 2017
K. Haastert-Talini et al. (eds.), *Modern Concepts of Peripheral Nerve Repair*,
DOI 10.1007/978-3-319-52319-4_8

exceeds 24 months, motor recovery will not be achieved due to irreversible muscle fibrosis. This limitation does not exist for sensory recovery, however, which can be expected even after delayed and high-level nerve repairs [19].

Patient age is another important factor that we must take into account. The results of nerve repairs in children are better, because of their higher potential for axonal growing and the shorter distances between the repair site and target muscle. The latency period between the lesion and the onset of reinnervation is also less. However, there is no distinct age threshold after which results suddenly become less favourable.

The mechanism of injury and severity of trauma also influence the outcome. Nerve injuries caused by traction have a worse prognosis than other injuries like stab wounds, because they affect large segments of the nerve. Associated bone fractures, vascular injuries and soft tissue defects, all indicators of trauma severity, generate ischemia, perineural scarring and/or defects in effectors, all of which negatively influence outcomes.

Certainly, the use of an appropriate surgical technique is essential to recovery. It is well known that the best results are obtained with end-to-end anastomoses; however, in most cases, using a nerve graft is necessary.

Graft length is another prognostic factor after the repair of large nerves (median, ulnar, sciatic, etc.). However, during repair of the brachial plexus, in an analysis conducted by the current authors, no differences were identified between using nerve transfers with long grafts (>10 cm) and the reconstruction of primary trunks with short grafts [24].

The importance of systematic rehabilitation is also well known. However, it is not easy to quantify adherence to rehabilitation when the results of nerve repair are analysed. The current authors have developed a scale for this purpose (Table 8.1). Using this scale to analyse results in a series of patients with brachial plexus injuries treated with long-graft nerve transfers, the fundamental role played by rehabilitation in determining outcomes was evident [24].

Furthermore, functional outcomes repairing different nerves in comparable circumstances are not always the same. Numerous studies have been published describing better results after the repair of the radial nerve than the median or ulnar nerves; there also is a better prognosis repairing the tibial versus peroneal nerve [11, 14].

To put treatment results in proper perspective, it is important to document these aforementioned factors. This is especially important for interpreting and evaluating the results of a specific treatment and, thereby, optimizing treatment strategies.

Table 8.1 The rehabilitation scale [24]

Score	Description
1	No rehabilitation at all or less than once per week
2	Rehabilitation more than once weekly, but not at a specialized centre
3	Good progress with a comprehensive rehabilitation programme, but not in a specialized centre; periodically supervised by a specialized neurorehabilitation centre
4	Patient adheres perfectly to the whole rehabilitation programme at a specialized neurorehabilitation centre

The degree of sensory and motor recovery is the criterion most commonly used to evaluate the results of nerve repair. Sensory recovery is not a reliable sign of regeneration, however, mainly because of its late appearance and difficulties with its objective evaluation. However, it is important after the repair of particular nerves like the median, ulnar and tibial nerve, as these provide protective sensation. It is less important as an outcome following the repair of nerves like the radial, axillary, musculocutaneous, femoral and fibular nerve.

On the other hand, although motor recovery is also late, it is a reliable sign of successful regeneration. It takes between 2 and 3 years to achieve maximum motor recovery, versus 5–7 years for maximum sensory restoration.

In general terms, the follow-up of any patient submitted to a peripheral nerve reconstructive surgery should be every 2 or 3 months. Clinical evaluation – including progression of Tinel's sign – and serial neurophysiological studies can help determine early recoveries. The regular endpoint of follow-up is around 3 years for motor results and around five for sensory recovery. At that time, it is presumed that the maximum recovery point will be reached. Of course it is not possible to generalize these time spans for every patient: depending on the time from trauma to surgery, the distance from the injury site to the target muscles/skin, the type of reconstruction and so on, the end of follow-up variates from patient to patient.

One important concept to keep in mind when analysing the results of nerve repair is that of 'useful recovery', which entails the functional impact of recovery. The definition of 'useful' is variable and depends on the nerve involved. For example, useful sensory recovery for the tibial nerve means recovery of superficial pain and some tactile sensation ($\geq$S2). However, for the median or ulnar nerve, it is also imperative to recover some two-point discrimination.

Similarly, useful motor recovery ($\geq$M3) after peroneal nerve repair involves plantar dorsiflexion to 90°, since this is enough for the patient to stop using a foot brace [19].

A variety of scales and questionnaires have been developed and published to objectify these results. They may be categorized according to the function they evaluate: motor, sensory, pain or global.

8.1 Sensorimotor Evaluations

The scale most commonly used to assess sensorimotor reinnervation is the British Medical Research Council (BMRC) scale [15] (Tables 8.2 and 8.3). This scale was promoted and standardized after the Second World War in order to eliminate or

Table 8.2 British Medical Research Council motor scale [12]

M0	No contractions
M1	Visible or palpable contractions
M2	Active movement, with gravity eliminated
M3	Active movement against gravity
M4	Active movement against resistance
M5	Normal power

Table 8.3 British Medical Research Council sensory scale [12]

S0	Absence of sensation
S1	Recovery of deep cutaneous pain sensation
S2	Recovery of superficial pain and some tactile sensation
S2+	Same as S2, with overresponse
S3	Recovery of pain and tactile sensation, with disappearance of overresponse
S3+	Same as S3, with some two-point discrimination
S4	Complete recovery

reduce interobserver variability. The motor and sensory components are not integrated, actually operating as two separate scales. The widely used motor scale can be used to evaluate either bulky muscles or muscle groups or small muscles like the intrinsic muscles of the hand. On the other hand, the sensory scale has been criticized [20] as being based on subjective parameters.

Tools exist to help us to quantify these results. The goniometer is used to evaluate active range of motion, and the dynamometer to measure muscle strength. The tested limb should always be compared with the contralateral limb (if normal).

The grip strength test (using a Jamar dynamometer) is the method most often used to communicate motor strength results.

Similarly, there are tools to quantify sensory results. The Semmes-Weinstein monofilament test can be employed to assess cutaneous pressure threshold. Compared to using a classical tuning fork, this test provides quantitative data that can be used to monitor nerve regeneration.

The two-point discrimination (2PD) test is a tool for evaluating tactile gnosis. *Tactile gnosis* is the hand's ability to recognize the characteristics of different objects, like their shape and texture. It is an important marker of functional recovery. One major flaw this test has is that the results can be variable, since there is no standardization of the technique, and different examiners perform the test differently. For this reason, it is important to provide a detailed description of the protocol used, especially the pressure applied. It also is not recommended that this test be used as the sole instrument to measure sensory function.

The shape/texture identification (STI) is a test developed by Rosen and Lundborg that consist of identifying three forms and three textures, with the index finger in median nerve injuries and the little finger if the ulnar nerve is affected. In patients with injury to both nerves, the index finger is assessed.

The Sollerman hand function test consists of 20 activities that replicate the main handgrips utilized in daily life, such as taking coins from a purse or undoing buttons. Each subtest has a score depending on the quality of handgrip and the difficulty the patient has performing the task. This test reflects the integration of sensory and motor functions.

8.2   Pain Evaluation

The numerical rating scale (NRS) for pain and the pain visual analogue scale (PVAS) are often used to determine pain intensity. Both are easy to apply but have flaws. First, they treat pain as a linear and continuous phenomenon, which is not so in most cases. On the other hand, not all patients respond to these scales the same way, since the experience of pain is very variable.

The McGill Pain Questionnaire [16] is a multidimensional scale that provides information not only on the intensity of pain but on other characteristics like sensation quality (e.g. sharp, pins and needles) and the patient's emotional response to pain. However, it is too long a questionnaire to readily integrate into daily clinical practice.

The Integrated Pain Score scale [2] (Table 8.4) allows us to record, over time, the characteristics and intensity of pain. Since it is simple and quick, it allows

Table 8.4   Integrated pain score [2]

Parameter	Description	PTS	Sum
Intensity (VAS)		0–10	✓
Incapacity		0–10	✓
Frequency of pain	Never	0	✓
	Rarely	1	
	Once a day	2	
	More than once a day	3	
	Continuous	4	
Use of pain medication	Never	0	✓
	Occasionally	1	
	Once a day	2	
	More than once a day	3	
	All the time	4	
	No alleviation	1	✓
Zones affected by pain	Distal	1	✓
	Medial	1	✓
	Proximal	1	✓
Sleep	Normal	0	✓
	Awakens only some nights	1	
	Awakens once every night	2	
	Awakens more than once every night	3	
	Insomnia	4	
	Use of hypnotics	1	✓
Total			✓

patients to be monitored successively before and after surgery. It separately analyses pain intensity and frequency, degree of disability, the use of analgesics, number of territories involved (proximal, middle or distal) and the effects of pain on sleep.

8.3 Global Scales

The Rosen scale was developed to allow for the documentation and quantification of functional outcomes after nerve repair at the wrist or distal forearm [21]. It includes three domains: sensory, motor and pain/discomfort. Motor function is assessed using the MRC scale and grip strength test (with a Jamar dynamometer). The evaluation of sensory function employs Semmes-Weinstein monofilaments, the 2PD to evaluate tactile gnosis, and the STI test. Pain/discomfort is evaluated using a scale with four grades (0–3) to categorize hyperesthesia and intolerance to cold.

The disability of arm, shoulder and hand (DASH) outcome questionnaire is an instrument developed specifically to assess results in the upper limb. It was introduced by the American Academy of Orthopaedic Surgeons in collaboration with other organizations. It is mostly a measure of disability. One of the objectives behind its development was to facilitate comparisons between different disorders affecting the upper limb. It is currently available in several different languages, including English, German, Italian, Spanish, Swedish, French and Dutch. It consists of a 30-item questionnaire addressing the patient's health status over the preceding week. Individual items ask about the patient's difficulty performing various physical activities due to problems in their shoulder, arm or hand (20 items); the severity of symptoms like pain, pain related to activity, tingling, weakness and stiffness (five items); and the impact these symptoms exert on social activity, work, sleep and self-image (four items). Each item has five possible answers. The final score ranges from 0 (no disability) to 100 (severe disability). According to a study by Gummesson et al. [9], this questionnaire can detect both small and large changes in disability over time post-operatively in patients with musculoskeletal disorders of the upper limb.

The Michigan Hand Outcomes Questionnaire (MHQ) consists of 37 items that evaluate disability across six domains: function, activities of daily living, pain, hand appearance, patient satisfaction and disability at work.

8.4 Post-operative Complications

8.4.1 Nerve Damage

Nerve damage during surgery on peripheral nerves is uncommon. The main cause of this type of unintentional injury is the surgeon's unfamiliarity with the regional anatomy accessed. There is also the possibility of so-called anatomic variants, which must be taken into account during any procedure, since failure to recognize such deviations from 'normal' can lead to the injury of such structures.

The anatomy of the lower limb has a lower percentage of anatomical variants than the upper limb. One example of an anatomical variant in the upper limb is the recurrent branch of the median nerve, which, though always present, can vary in its location and occasionally be damaged if not recognized.

The nerves most often injured are the median nerve, the ulnar nerve, the digital nerves and communicating branch between the median and ulnar nerve (Berretini's branch or anastomosis). Many nerves are susceptible to damage, like the palmar cutaneous branch and the motor branch of the median nerve, the external digital nerve to the fourth finger and the common digital nerve for the third and fourth finger [8].

Fibrosis of the median nerve, both intra- and perineural, can result from chronic compression and be secondary to surgery [29]. Fibrosis of this nerve frequently causes highly annoying dysaesthesias, intense local pain and local skin hypersensitivity. In these cases, neither internal nor external neurolysis is indicated, since it has not been demonstrated that good results are obtained with these techniques.

8.4.2 Vascular Injury

During virtually every surgery, an anatomical region with blood vessels is approached. The location of these vessels, as with major nerves, tends to be anatomically constant; however, collateral branches can vary significantly. Obviously, the severity of the vascular insult and resulting bleeding and ischemia depends upon the size and status of the vessel involved: injury to a major arterial branch is of much greater concern than injury to a fourth-order branch or distal vessel.

During surgery to peripheral nerves, we sometimes elect to 'forget' distal vessels that cross our surgical field, thereby rendering the nerve dissection more difficult. We can sometimes afford to do so thanks to collateral circulation that permits us to section such vessels without significant consequences. Before doing this, however, we have several things to determine, like whether or not collateral circulation actually exists; whether the vessel might be used in some other reconstruction procedure like a free vascularized graft; and whether the calibre and flow of the vessel are too great to allow for adequate spontaneous coagulation, so that it becomes necessary to use vascular clips to prevent post-operative bleeding [18].

Vascular injury may occur even during seemingly simple nerve surgeries, like carpal tunnel decompression. For example, if we extend the opening of the carpal ligament distally, we can injure the superficial and/or deep palmar arches that provide circulation to the fingers [25].

8.4.3 Postoperative Bleeding

Bleeding is a common complication of all types of surgery. In the case of nerve surgery, it is rare if all haemostatic measures are taken in each plane of dissection.

Compressive haematoma formation is highly unusual but can result from increased blood pressure post-operatively. Both blood pressure and the presence of haemorrhage must therefore be monitored closely; and if a local haematoma increases in volume to a point at which there is a risk of ischaemia with further enlargement (e.g. compartment syndrome), immediate intervention is necessary to stop the bleeding [27].

With mild bleeding, there is some risk of adhesion formation or flanges that can, in the long term, cause nerve compression associated with different symptoms like paraesthesia, hypoesthesia and pain.

8.4.4 Wound Complications

8.4.4.1 Wound Dehiscence

When planning any surgery, it is important to consider which skin incision to use, because it not only has an aesthetic but legal connotations, sometimes leading to legal claims.

It also should be considered during planning that previous scars might alter skin circulation. The patient's metabolic state is important too, since diabetes may alter normal healing, and patients with hypoproteinaemia take longer to heal their wounds. Another important issue is the presence of skin folds; for example, with carpal tunnel surgery, the skin incision should not cross the transverse folds of the wrist, as this can lead to the formation of a hypertrophic scar or keloid [7]. Six months after surgery, only 2% of patients still have tender scars. When a post-operative patient presents with a painful hypertrophic scar, surgical revision may become necessary, even performing a Z-plasty to avoid new hypertrophic scarring and associated discomfort.

8.4.4.2 Wound Infection

As with bleeding, wound infections are common to all types of surgery. The incidence of superficial skin infections reported in the literature is between 0.5% and 6% for carpal tunnel surgery. Deep infections have also been reported after surgery for carpal tunnel syndrome. Risks for deep infections include the use of drains, prolonged surgery and performing a synovectomy of carpal tunnel tendons [23].

8.4.5 Pain

The patient who presents with a traumatic peripheral nerve injury associated with pain must report certain features for the pain to actually be ascribed to the injury. In the first place, discomfort should be felt in the distribution of the affected nerve [26]. Additionally, the area should experience anaesthesia or sensory loss, as it is rather unlikely for someone to feel pain originating from an injured nerve without some tactile alterations, regardless of whether the anaesthesia is complete or partial. In general, pain is described as a burning sensation, as an electric shock or as a very annoying tingling. Quite often, pain is triggered by stimuli that would otherwise be non-painful, such as gentle touch (allodynia) [3, 4]. On clinical examination, there may be autonomic changes in the distribution of the nerve, albeit slight and unlike those of the reflex sympathetic dystrophy mentioned below. As well, the patient may have a positive Tinel's sign, which the examiner can elicit by percussing directly over the proximal end of the nerve. In doubtful cases, it may be useful to perform a nerve block with local anaesthetics to confirm the diagnosis [28].

The most important differential diagnoses for neuropathic pain are the complex regional pain syndromes (formerly known as reflex sympathetic dystrophy) [17, 30]. In these patients, pain initially is acute and dysaesthetic, often felt as a burning sensation, distributed distally, and associated with exaggerated responses of the sympathetic nervous system (e.g. sudden flushing and increased sweating immediately followed by pallor and limb coldness). Sympathetic blockade helps to confirm the diagnosis and initially provides a certain degree of therapeutic relief. Long-term control is difficult, however, because the pain tends to be chronic.

After it has been determined that the pain is due to an injured nerve, medical treatment should be administered. One of the most commonly used drugs (and preferred by the current authors) is gabapentin, starting at a daily dose of 600 mg, usually plateauing at about 1800 mg, but with a maximum allowable dose of 3600 mg daily. Amitriptyline, carbamazepine, and ultimately pregabalin are other drugs that are commonly used to treat neuropathic pain. Other alternatives, like non-steroidal anti-inflammatory drugs and electrostimulation [13], have proven useful in some cases.

With patients for whom medical treatment is insufficient to alleviate symptoms, there are two initial surgical options. First, if the nerve is small, superficial and predominantly sensory, as is the case with the sensory branch of the radial or sural nerve, it can be directly resected proximal to the injury. With sectioning of the nerve, the pain disappears at the price of cutaneous anaesthesia in the area supplied by the peripheral nerve sectioned; almost always, this is well tolerated. This section should be done, if possible, 15–20 cm proximal to the affected area [1]. On the other hand, if the nerve cannot be sacrificed, as with the greater sciatic nerve, then neurolysis or reconstruction with a graft, depending on the severity of the injury, is the treatment of choice [5]. The right time to perform any of these procedures depends on the pain's response to drugs. However, it is generally agreed that a traumatized nerve that generates pain should be surgically explored if it does not respond to pharmacological treatment within a reasonable time frame, not to exceed 45 days [12].

Conclusion

Clinical follow-up in peripheral nerve surgery greatly varies depending on the time and site of the primary injury, among many other factors that influence the outcome. Several scales exist, designed to measure the final outcome of a nerve reconstruction, which is of paramount importance when comparing different techniques. At present, there is not consensus towards this point, and different departments use a different way to determine their results. Hopefully, this problem will be solved in the future if a unique method to evaluate results is employed.

References

1. Bertelli J, Ghizoni MF. Pain after avulsion injuries and complete palsy of the brachial plexus: the possible role of nonavulsed roots in pain generation. Neurosurgery. 2008;62(5):1104–13.
2. Bonilla G, Di Masi G, Battaglia D, Otero JM, Socolovsky M. Pain and brachial plexus lesions: evaluation of initial outcomes after reconstructive microsurgery and validation of a new pain severity scale. Acta Neurochir. 2011;153:171–6.
3. Burchiel KJ, Johans TJ, Ochoa J. Painful nerve injuries: bridging the gap between basic neuroscience and neurosurgical treatment. Acta Neurochir Suppl (Wien). 1993;58:131–5.
4. Burchiel KJ, Johans TJ, Ochoa J. The surgical treatment of painful traumatic neuromas. J Neurosurg. 1993;78(5):714–9.
5. Burchiel KJ, Ochoa JL. Surgical management of post-traumatic neuropathic pain. Neurosurg Clin N Am. 1991;2(1):117–26.
6. Dong Z, Zhang CG, Gu YD. Surgical outcome of phrenic nerve transfer to the anterior division of the upper trunk in treating brachial plexus avulsion. J Neurosurg. 2010;112:383–5.
7. Ferraresi S. Actas del I Congreso Chino-Europeo de Cirugía del Plexo Braquial. Venecia, marzo; 2008.
8. Friedman AH, Nashold Jr BS, Bronec PR. Dorsal root entry zone lesions for the treatment of brachial plexus avulsion injuries: a follow-up study. Neurosurgery. 1988;22(2):369–73.
9. Gummesson C, Atroshi I, Ekdahl C. The disabilities of the arm, shoulder and hand (DASH) outcome questionnaire: longitudinal construct validity and measuring self-rated health change after surgery. BMC Musculoskelet Disord. 2003;4:11.
10. Jivan S, Kumar N, Wiberg M, Kay S. The influence of presurgical delay on functional outcome after reconstruction of brachial plexus injuries. J Plast Reconstr Aesthet Surg. 2009;62:472–9.
11. Kline DG. Operative management of major nerve lesions of the lower extremity. Surg Clin North Am. 1972;52:1247–65.
12. Kline DG, Tasker RR. Pain of nerve origin. In: Kim DH, Midha R, Murovic JA, Spinner RJ, editors. Kline and Hudson's nerve injuries. 2nd ed. Nueva York: Saunders; 2008. p. 415–26.
13. Laryea J, Schon L, Belzberg A. Peripheral nerve stimulators for pain control. Semin Neurosurg. 2001;12(1):125–31.
14. Mackinnon SE, Dellon AL. Results of nerve repair and grafting. In: Surgery of the peripheral nerve. New York: Thieme; 1988. p. 123–4.
15. Medical Research Council (MRC) Nerve Injuries Committee. Results of nerve suture. In: Seddon HJ, editor. Peripheral nerve injuries. London: Her Majesty's Stationery Office; 1954.
16. Melzack R. The McGill Pain Questionnaire: major properties and scoring methods. Pain. 1975;1(3):277–99.
17. Melzack R, Wall PD. Pain mechanism: a new theory. Science. 1965;150:971–9.
18. Midha R, Zager EL. Surgery of the peripheral nerves: a case-based approach. 1st ed. New York: Thieme; 2008.
19. Roganoviæ Z. Repair of traumatic peripheral nerve lesions: operative outcome. In: Siqueira MG, Socolovsky M, Malessy M, Indira Devi B, editors. Treatment of peripheral nerve lesions. Bangalore: Prism Books PVT Ltd.; 2011. p. p11–120.
20. Rosén B. Recovery of sensory and motor function after nerve repair. A rationale for evaluation. J Hand Ther. 1996;9(4):315–27.
21. Rosén B, Lundborg G. A model instrument for the documentation of outcome after nerve repair. J Hand Surg Am. 2000;25(3):535–43.
22. Samii A, Carvalho GA, Samii M. Brachial plexus injury: factors affecting functional outcome in spinal accessory nerve transfer for the restoration of elbow flexion. J Neurosurg. 2003;98:307–12.
23. Socolovsky M. Conceptos actuales en la cirugía de los nervios periféricos: Parte I: Lesiones del Plexo Braquial. Rev Arg Neurocir. 2003;17(2):53–8.

24. Socolovsky M, Di Masi G, Battaglia D. Use of long autologous nerve grafts in brachial plexus reconstruction: factors that affect the outcome. Acta Neurochir. 2011;153:2231–40.
25. Socolovsky M, Di Masi G, Campero A. Lesiones de los nervios periféricos: ¿Cuándo operar? Rev Arg Neurocir. 2007;20(2):71–9.
26. Sunderland S. Nerve injuries and their repair: a critical appraisal. London: Churchill Livingstone; 1991.
27. Taggart M. Rehabilitation. In: Birch R, Bonney G, Wynn Parry CB, editors. Surgical disorders of the peripheral nerves. 1st ed. Edimburg: Churchill Livingstone; 1998. p. 461–3.
28. Tasker R, De Carvallo GTC, Dolan EJ. Intractable pain of spinal cord origin: clinical features and implications of surgery. J Neurosurg. 1992;77:373–8.
29. Terzis JK, Papakonstantinou KC. The surgical treatment of brachial plexus injuries in adults. Plast Reconstr Surg. 2000;106:1097–122.
30. Wynn Parry CB. Pain in avulsion of the brachial plexus. Neurosurgery. 1989;15:960–5.

Rehabilitation Following Peripheral Nerve Injury

9

Sarah G. Ewald and Vera Beckmann-Fries

Contents

9.1 Introduction

The impact of a peripheral nerve injury (PNI) on an individual's function ranges from moderate and temporary to significant and life changing, depending upon the severity and location of the injury and patient-specific factors. The functional loss that results from a PNI can be sensory, motor or sensory, and motor in nature. The quality of life (QoL) for these patients is substantially reduced with approximately 25% of patients still out of the workforce 1.5 years after surgery (as cited by Davis et al. [5]). Not only do patients lose function, unfortunately they may also gain

S.G. Ewald, OTR, MA Ed (✉)
City Handtherapie, Stampfenbachstrasse 42,
CH-8006 Zurich, Switzerland
e-mail: sgewald@city-handtherapie.ch

V. Beckmann-Fries, PT, MME
UniversitätsSpital Zürich, Physiotherapie Ergotherapie/Handtherapie,
Gloriastrasse 25, CH-8091 Zurich, Switzerland
e-mail: vera.beckmann-fries@usz.ch

© Springer International Publishing AG 2017
K. Haastert-Talini et al. (eds.), Modern Concepts of Peripheral Nerve Repair,
DOI 10.1007/978-3-319-52319-4_9

function in the form of persistent pain, hyperesthesia, hyperalgesia, or allodynia. Rehabilitation following injury to a peripheral nerve is essential to optimize function and outcomes. Clinical studies have demonstrated that patients that receive hand therapy tailored to their needs have better outcomes, for example, for sensory function, than patients that do not receive treatment [3, 17]. Rehabilitation of PNI in the upper extremity is performed by occupational and physical therapists; many of these therapists have the additional qualification of "hand therapist," which indicates additional training in this specialty practice area of both professions. Worldwide, there are about 8500 therapists that practice "hand therapy" [9]. The hand therapist tailors the treatment in response to the functional loss that the patient presents with. Not only functional loss will be addressed in therapy. The so-called gain in function, mainly in sensory input as hyperesthesia, hyperalgesia, or allodynia, will be addressed in therapy as well. Thus, the course of therapy may be highly variable between patients: some need functional therapy and intense sensory rehabilitation, and for others afflicted with intense pain, pain treatment and education about pain management will be given a greater priority in therapy.

9.1.1 Site/Level of Injury

In general, the more proximal the injury, the greater the impact is on function. A brachial plexus injury is a severe and devastating injury requiring in most cases multiple surgeries and a prolonged course of rehabilitation. A more distal injury at the level of the wrist still impacts function significantly and may have life-changing implications. For example, an injury resulting in a laceration of the median and ulnar nerves at the level of the wrist results in loss of sensation and partial loss of motor function in the hand. Loss of sensation in the hand significantly impacts hand dexterity and overall hand function and can result in further injury to the hand due to the lack of protective sensation. In comparison, laceration of the radial nerve at the level of the wrist results in the lack of sensation on the dorsum of the hand, and motor function remains intact. As the dorsum of the hand is less directly involved in tasks that require dexterity, the loss of this sensation, while surely an irritant, has less impact on overall function. However, laceration of the sensory branch of the radial nerve can develop as possible source of persistent pain [8].

This chapter will focus on the rehabilitation process for peripheral nerve injuries that occur distal to the brachial plexus/mid-humerus in the upper extremity.

9.2 Rehabilitation

Following PNI, it is recommended that a patient be referred as early as possible for therapy [17, 26]. The rehabilitation program is designed and implemented by the therapist after careful evaluation of the patient and his/her condition and functional level. As the patient progresses, evaluation is ongoing, and treatment is continuously adapted and modified in response to evaluation results; this occurs in all phases of rehabilitation.

9.2.1 Evaluation Methods

The therapist's assessment of the patient is comprised of a variety of evaluation methods and tools. The multifactorial evaluation helps the therapist gain a clear picture of the functional impact of the injury and the patient's needs. It also provides a baseline against which progress or lack thereof can be measured. An overview of evaluation tools frequently utilized by hand therapists when assessing patients with PNI can be found in Table 9.1. The results of the evaluation guide the development of the therapy program and often motivate the patient to continue with exercises. The therapist evaluates not only motor and sensory function as components of body function but also the patient's activity and participation levels. Incorporating the International Classification of Functioning, Disability and Health (ICF) with its bio-psychosocial model and classification system into the evaluation process is advantageous as it offers all members of the health-care team a common framework when viewing the patient and his or her individual situation.

For example, evaluation of range of motion, strength, and sensory function certainly help to determine levels of biomechanical function and impairments, but this does not translate into an assessment of overall functional status in multiple life domains of the patient. It is important to consider the psychosocial factors such as coping, depression, and anxiety as well. The rehabilitation process should also focus on improvements in quality of life (QoL) and function in daily living. There is a "need for increased screening and assessment of factors that reach beyond the biomedical model" [28]. Figure 9.1 illustrates the multiple factors that can impact the outcome for a patient with a PNI. While many of the factors are a given such as age and cognitive capacity, some of them can be influenced during the rehabilitation process.

A multicentered prospective study in the Netherlands [11] found that sensibility of the hand, grip strength, and the disabilities of arm, hand, and shoulder (DASH) questionnaire score were the best prognostic factors for functional recovery after peripheral nerve injury (median and/or ulnar nerve) at the level of the forearm. The Rosén Score, developed for median and ulnar nerve injuries at the level of the forearm, is a helpful prognostic tool for surgeons, therapists, and patients [20, 21, 25, 27].

The therapist's evaluation is ongoing and essential for her clinical reasoning processes and planning therapy. Once the initial evaluation has been completed, treatment is implemented. The treatment plan is customized to the needs of the patient, and the therapist will make use of many tools, methods, and approaches in an effort to treat the patient at a very individual level.

Table 9.1 Evaluation tools used by therapists

| To be evaluated | Evaluation tools | | | |
| | Objective measure | | Subjective measure | |
	Test	Prognostic tool	Rating scales	Questionnaires
Pain and cold sensitivity		Rosén Score – a tool to assess functioning and pain of patients suffering from median and/or ulnar nerve laceration at wrist or distal forearm level	Visual analog scale (VAS)	McGill Pain Questionnaire
			Numeric rating scale (NRS)	Cold Sensitivity Severity (CSS) Cold Intolerance Symptom Severity (CISS)
Motor function	Reflexes Manual muscle testing Grip and pinch strength testing with dynamometer and pinch gauge			
Sensory function	Two-point discrimination (2PD) Semmes-Weinstein monofilament testing (SWM) Modified Moberg pickup test Shape/texture identification (STI) test Grating domes Localization of touch (as locognosia test)		Ten test	
Functional use of hand requires motor and sensory function	Activities of daily living (ADL) checklist 9-Hole Peg Test Functional Dexterity Test (FDT) Purdue Pegboard Test Jebsen Taylor Test of Hand Function Sollerman hand function test			Disabilities of arm, hand, and shoulder (DASH) or *Quick*DASH Canadian Occupational Performance Measure (COPM) Michigan Hand Questionnaire (MHQ)
Quality of life				Short Form 36 (SF-36)
Sense of coherence				Antonovsky's short 13-item scale

Biological Factors

- Age
- Extent of Injury (additional structures involved)
- Cognitive Capacity and Function, etc.

Individual Factors

- Personality
- Motivation
- Sense of Coherence
- Goal Orientation
- Adherence to Therapy, etc.

Environmental Factors

- Therapy
- Life Situation
- Family and social support, etc.

Fig. 9.1 Influences on outcome after PNI

9.2.2 Rehabilitation: Considerations and Methods

The therapeutic approach is multimodal and may include treatment to decrease pain and edema, the use of custom-made or prefabricated splints, instruction in adaptive methods to improve function in day-to-day life, and facilitation of motor relearning and sensory reeducation. Nerves recover slowly after laceration and repair, notably at the rate of 1–3 mm per day after a 2–3-week latency period [6], and a prolonged course of therapy is usually required. The frequency of therapy may vary considerably depending on the nerve that is injured and the subsequent impact on the patients' function and the phase of recovery. The focus of therapy in the initial phase, defined as the early postoperative phase, is on protecting the injured structures, reduction of postoperative swelling, pain management, maintaining function in adjacent noninjured structures, preserving cortical representation, and improving the level of function in daily activities.

As the patient progresses to the innervation phase, therapy focuses on regaining motor and sensory function and reintegrating this function into the overall function of daily living. Pain management or addressing sensory gain of function as hyperesthesia may still be a treatment aim. The patient may require splints that balance or support muscle function in the hand. At each step of the recovery process, it is important that patients are given tools and instructions that allow them to participate fully in the process and facilitate their recovery. A well-designed home program is essential at every step of the rehabilitation process. The program should be reevaluated and adjusted at regular intervals. Effective rehabilitation makes use of the patients' resources and abilities during each of the recovery phases. It is beneficial for a patient to attend therapy at regular intervals to monitor progress and adapt the home program. For some patients, once the home program is established, it may be sufficient for therapy visits to take place on a weekly or even monthly basis. Recovery may be prolonged, and periods of intensive therapy are required when treatment is initiated and periodically in response to changes or additional surgery. In this section, we will describe some of the tools and methods used in therapy to achieve these aims.

9.2.2.1 Postoperative Swelling and Protection of the Sutured Nerve

Postoperative swelling is to be expected and is treated much the same as for any other injury, that is, with positioning, elevation, manual edema mobilization, and compression. An initial period of protection following the operative repair of the nerve is recommended [6, 22]. In the initial phase of therapy, therapists protect the repaired nerve with splinting or bandaging. The surgeon in charge defines the period of protection; it depends on the location of the injury and how much tension there is on the repair site. While it is important to protect the repair, to limit the expected side effects on non-involved tissue and joints, it is advisable to immobilize the fewest joints and soft tissues possible without compromising the repair [14].

9.2.2.2 Pain Management

"Pain has been defined as a multidimensional experience consisting of sensory-discriminative, affective–motivational, and cognitive–evaluative components" [24]. After peripheral nerve injury, pain is to be expected, and it is a priority to address all aspects of pain as early as possible in the course of treatment. Strategies for pain management are manifold, as shown in Table 9.2.

Some patients develop neuropathic pain after peripheral nerve injury. Pain management for these patients is of utmost importance and includes the previously mentioned strategies; additional psychological support must be considered. It is recognized that neuropathic pain is associated with a poor outcome and high levels of disability [16]. Patient outcomes, such as level of pain, disability, and patient satisfaction, are influenced by psychosocial factors, such as depression, coping, and anxiety [28]. It is therefore paramount that the therapists assess and address these psychosocial factors as part of the rehabilitation process.

Table 9.2 Overview of pain management methods used in hand therapy following peripheral nerve injury

Management used	Aim
Positioning the affected body part	Prevention of secondary harm through mal positioning
Sensory input	Providing positive, comfortable sensations
Thermal modalities	Decrease of muscle tone proximal to the injured side and/or positive comfortable sensations
Transcutaneous electrical nerve stimulation (TENS)	Triggers the gate and opiate systems for pain modulation
Graded motor imagery (which includes mirror therapy)	Maintain and/or reawaken cortical representation
Providing information about injury, pain mechanism, coping strategies	Improve patient's ability to cope with and self-manage pain

9.2.2.3 Scar Optimization

As with any other type of injury, it is important that the scar be as mobile, flexible, and esthetically appealing as possible. To this end, the therapist may employ scar massage, taping, electrophysical modalities, and in some cases the use of silicone with compression dressings [2]. Patients can be instructed to perform daily scar massage as part of their home program. The use of silicone dressings for scar treatment is to be avoided if there is a problem with gain of function: hyperesthesia, hyperalgesia, or allodynia. In our experience, the silicone dressing is often so comfortable for the patient that as sensation returns, which is often quite uncomfortable, the patient may refuse to discontinue the use of the dressing and the hypersensitive area thus becomes even more uncomfortable.

9.2.2.4 Improving Function and Activities of Daily Living

Gaining functional use of the extremity and overall function in day-to-day life is ongoing. It is important that the extremity be integrated into any activity as much as possible. This may mean the injured extremity can only be used initially to stabilize objects, while the uninjured hand performs the majority of the task. Maintaining cortical representation for the injured extremity is essential, the brain is plastic and remodels rapidly, and nonuse of the extremity will result in a cortical reorganization [15], which may be challenging to remodel as recovery of the nerve and function progresses. Not only cortical representation is very important, therapist must also consider the mobility of torso and shoulder; if the upper extremity is not incorporated into daily tasks, the flexibility of the thoracic spine diminishes rapidly, resulting in unwanted side effects.

During therapy, the patient will be instructed in adaptive methods, and adaptive equipment will be provided as needed to accomplish activities of daily living. As patients' needs vary depending on their living, work, and family situations as well as their coping mechanisms, this process is highly individualized. The use of ADL checklists and questionnaires such as the DASH can be very helpful in identifying areas of function that can be improved upon with therapy.

9.2.2.5 Sensory Function and Reeducation

Initially, a nerve laceration results in an absence of sensation. When there is no sensation present, it is easy to burn or cut the hand in the area that lacks sensation. Patients must be taught to inspect the body area that lacks sensation on a daily basis and to care for any injuries to this area carefully. It is important to use the hand to maintain integration in body schema. The use of a body part that has no sensation is quite challenging, it requires close visual control on the part of the patient, and this is a skill that must be learned.

As the nerve recovers and sensory function resumes, in some cases, the patient may experience an excess of sensation, so-called hyperesthesia, in the area where sensation is recovering. This can be quite uncomfortable and result in protective behavior and nonuse of the hand. In this situation, the therapist will instruct the patient in a desensitization program. Desensitization uses materials, contact particles, and vibration to reduce the hyperesthesia of the affected body part [4, 29]. Frequent and regular application of stimulus, beginning with material that is just tolerable to touch and progressing toward materials that are perceived to be more noxious, is required to achieve the desired effect. Significant improvement in the level of discomfort resulting from the hyperesthesia has been observed with a 6-week course of desensitization treatment [4, 10, 29]. Patients are instructed in a home program that is monitored closely and adjusted regularly. Most patients see an improvement within 1–2 weeks, although it may take several weeks of treatment with the home program until feelings of hyperesthesia are sufficiently resolved to allow the hand to be used freely.

It is critical to distinguish between hyperesthesia and allodynia. In the case of allodynia, stimulation of the affected area must be avoided; classical desensitization will exacerbate the perceived pain, not reduce it. Allodynia is part of neuropathic pain; a different approach such as "somatosensory rehabilitation" as described by Spicher [23] should be utilized.

Sensory reeducation following nerve injury must start as early as possible. In patients for whom sensory reeducation was initiated early, discriminative sensation was significantly better at 6 months than in patients for whom sensory training was initiated when some sensory function had recovered [17, 19]. In the early phase when no sensation is present, sensory reeducation may take the form of mirror therapy. Mirror therapy utilizes a mirror, and the patient placed with the mirror perpendicular at midline so that the injured hand is hidden from view behind the mirror and the uninjured hand is viewed in the mirror and reflected as if it were the uninjured hand. Rosén et al. [17] describe the following exercises for early sensory reeducation: initially the patient is instructed to view the uninjured hand in the mirror, then to name the fingers on the hand while viewing the hand in the mirror, then to slowly move the uninjured hand, and then to move both hands symmetrically. A partner can be involved and asked to symmetrically gently touch both hands while the patient observes in the mirror. It is thought that mirror therapy helps to maintain cortical representation of the affected body area (Fig. 9.2).

Early sensory reeducation includes the use of the sensible hand for activity. Patients should be encouraged to use their visual and auditory capabilities to process information that is encountered by the hand. Two-handed activities, such as shoe tying, peeling an orange, and opening a packet, should be encouraged when motor function is present. When no motor function is present, the hand can assist when positioned to stabilize objects during activity. These methods help maintain cortical representation and integration of the affected area.

The perception of moving touch recovers earlier than static touch. As moving touch begins to recover, the patient can be instructed in a simple home program that can be done with a partner. The partner is instructed to use the tip of a pen to "draw" different types of lines along the affected area. The patient is asked to identify the type of line.

The next step in the course of sensory reeducation is made when the patient is able to discern Semmes-Weinstein monofilament 4.31 or 4.56 [17]. With this level

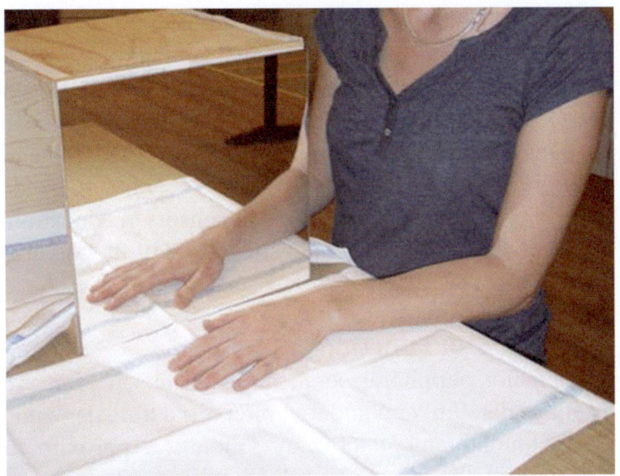

Fig. 9.2 Mirror Therapy: The patient is asked to focus his attention on the image in the mirror. The mirror reflects the uninjured hand as the contralateral hand. The use of a mirror in this manner can be helpful for pain reduction and is used in early sensory reeducation

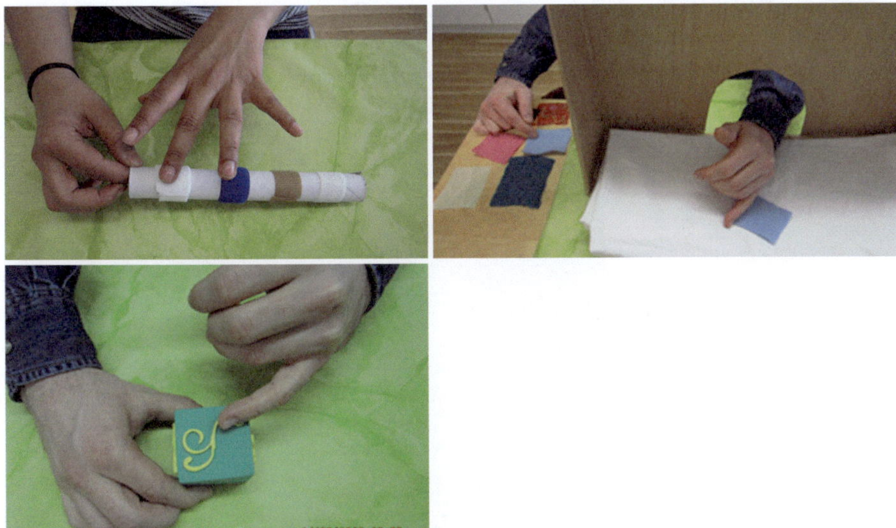

Figs. 9.3, 9.4, and 9.5 Image 3 Sensory Roll: Patient discriminates between four different surfaces of equal width. Image 4 identifying different types of materials. Image 5 discriminating among different raised shapes on a sensory block

of sensation present, it is possible to begin more formal sensory retraining that focuses on the identification and discrimination of different surfaces and forms. Examples of these types of exercises are visible in Figs. 9.3, 9.4, and 9.5. In this phase, it is essential to start with a small number of items and surfaces that can be identified by the patient. These items and surfaces should be significantly different from one another so that the patient has success. As the patient's speed and accuracy with identification improves, the difference in the surfaces and objects should lessen. The objects and surfaces that are part of the sensory reeducation program are refined continuously to challenge but also at a pace that allows the patient to succeed in completing the task. Sensory reeducation is challenging and requires intense concentration on the part of the patient. It is recommended that a patient has a home program and the training take place several times a day for short periods of time (5–10 min). Recent innovations in sensory reeducation of the hand include the use of selective temporary anesthesia at level of the forearm to enhance sensation in the injured hand [13, 18].

9.2.2.6 Motor Function

After major peripheral nerve injuries, muscles innervated by the involved nerves atrophy (Fig. 9.6) and undergo interstitial fibrosis, with initial weight loss of 30% in the first month and 50–60% by 2 months. Approximately 4 months after laceration and surgery, the muscle atrophy reaches a relatively stable state at 60–80% weight loss. The likelihood of functional reinnervation of the affected muscles diminishes within about 12 months. This is a consequent of the progressing fibrosis [12]. The initial phase of treatment focuses on instructing the patient in passive range of motion exercises to be performed daily to preserve joint mobility and prevent contractures. As nerve function recovers and the muscle is reinnervated, active

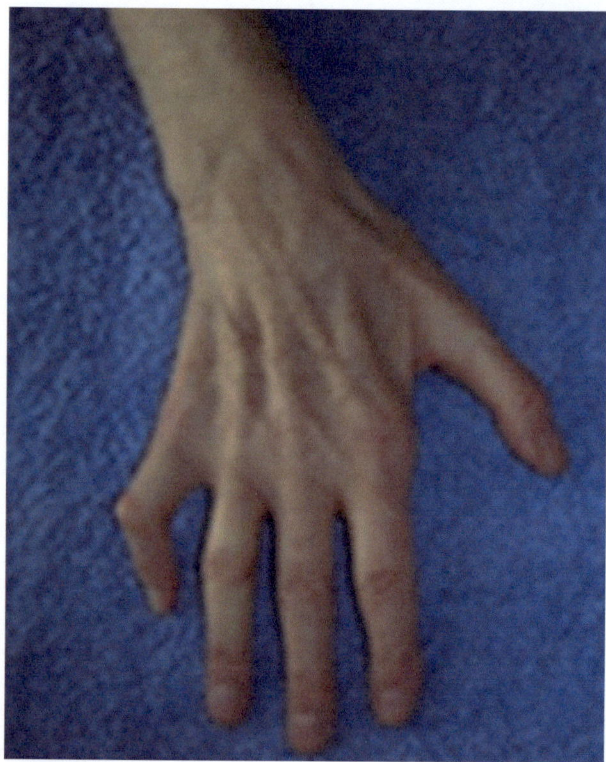

Fig. 9.6 Intrinsic muscle atrophy and contracture of fourth and fifth finger following ulnar nerve injury

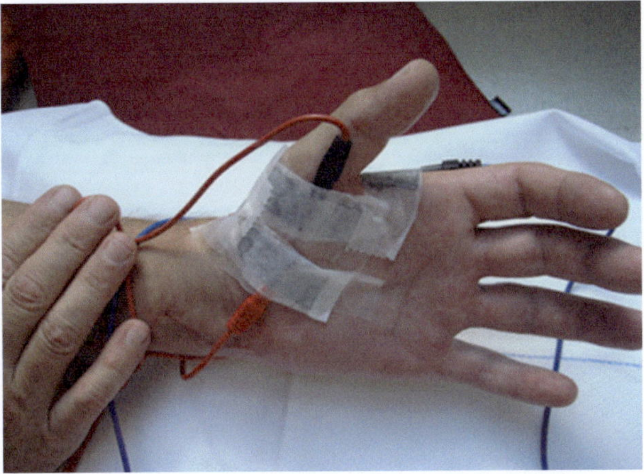

Fig. 9.7 Electrical stimulation of denervated muscles (post laceration of N. ulnaris at wrist level)

exercises to encourage muscle function should be instructed, progressing from gravity eliminated planes to resistive exercises that facilitate the development of muscle strength and control. In some clinics, functional electrical stimulation (Fig. 9.7) is used to maintain motor function in the absence of nerve function.

The electrical stimulation may prevent or delay some degree of muscle atrophy [1]. However, the level of evidence in clinical studies for electrical stimulation of denervated muscle is limited and based on small case series reports [16]. In our experience, patients like this type of treatment, and the visual aspect of this therapy should not be underestimated. The patient observes the muscles contracting and actively experiences these movements. Maintaining or restoring normal motor patterns is an important treatment goal. During the interval from of time from injury to reinnervation, many patients develop compensatory movement patterns [16]. It is sometimes quite a challenge to restore normal movement patterns once the compensatory movement has been established. Motor function is easier to implement in everyday purposeful activity – but this requires functional sensibility. If the patient has no sensation in his affected fingers, it is unlikely that he will use the affected body part and integrate the reinnervated muscle functions. Motor functions underlie the same principles of cortical reorganization as sensory functions. Elbert and Rockstroh [7] coined the following phrases that we find useful when educating patients about motor recovery: "practice makes perfect; use it or lose it; fire together, wire together; you have to dream to achieve it."

Recovering motor function is often supported with the use of custom-made or prefabricated splints.

9.2.2.7 Splinting

Splints are used to protect a structure, prevent contractures, correct deformities, enhance movement, and/or facilitate function. As splints to protect a structure can be easily imagined, we will focus here on splints that enhance movement, facilitate function, and prevent or correct deformities. Therapists are often challenged to

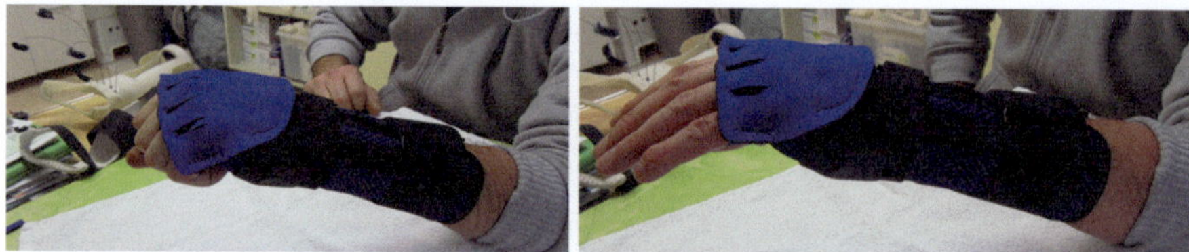

Fig. 9.8 Low profile splint for radial nerve injury

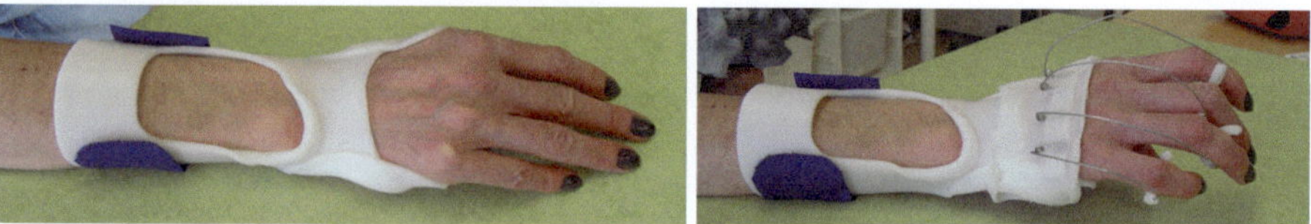

Fig. 9.9 Radial nerve splint with removal dynamic outrigger for finger extension

create a splinting solution that enhances function and is comfortable and acceptable to the patient.

Splinting for Radial Nerve Injuries

Injury to the radial nerve above the elbow results in paralysis of the extensors of the wrist and fingers, and as a result the flexors are unopposed. Finger flexion is possible but is now accompanied by wrist flexion as the wrist extensors are not working. Grasping an object is possible but releasing the object with the lack of finger extensor function becomes difficult. For smaller objects, it may suffice to relax the flexors to release an object, but for larger objects, the lack of finger extensor muscle function will prevent release of the object. The function of the extensor muscles can be replicated with splints. Both custom-made and prefabricated splints can be used to replicate extensor function and simultaneously prevent the flexor muscles from shortening and the extensor muscles from overlengthening during the recovery period. Some examples of splints used for radial nerve injuries can be seen in Figs. 9.8 and 9.9.

Splinting for Ulnar Nerve Injuries

In the case of an ulnar nerve injury, the intrinsic muscles of the hand are affected. When intrinsic muscle function is absent, the extensor muscles have no antagonist, and this results in hyperextension of metacarpal phalangeal (MCP) joints and flexion of the interphalangeal (IP) joints of fourth and fifth fingers. To balance the pull of the extrinsic extensor muscles on the MCP joints, a simple splint can be constructed that guides the MCP joints of the fourth and fifth finger into flexion and provides resistance to the extensors. The hand becomes much more functional with the use of such a splint, and flexion contractures of the interphalangeal joints are

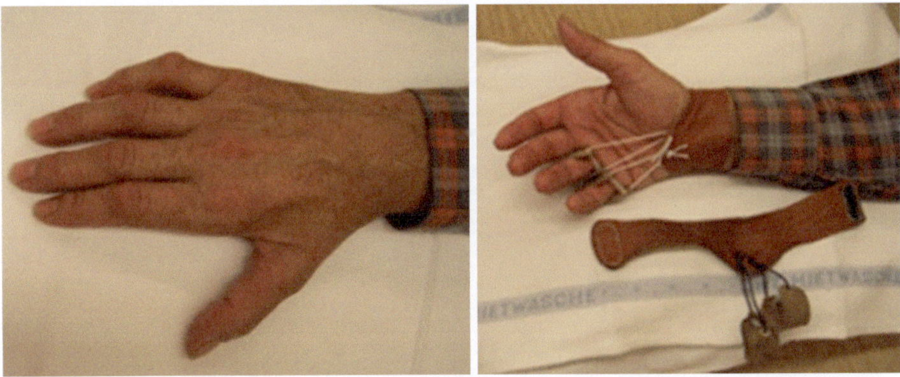

Fig. 9.10 Soft splint for ulnar nerve injury

Fig. 9.11 Thermoplastic splint for ulnar nerve injury

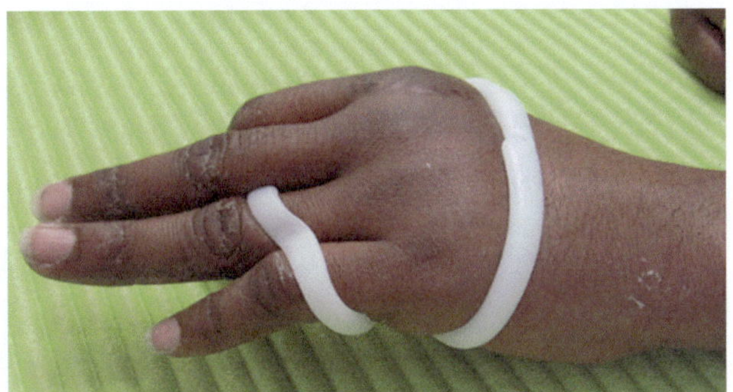

prevented. In our experience, patient acceptance of such splints is very good, particularly when they are permitted some choice as to color and materials and are fabricated to be as unobtrusive as possible. Example of splints used for ulnar nerve injuries can be seen in Figs. 9.10 and 9.11.

Splinting for Median Nerve Injuries

When the median nerve is injured, for example, at the level of the wrist, the intrinsic muscles of the thumb and to some extent the index and long fingers (lumbrical muscles) are impacted. This results in decreased thumb function as opposition is lost, as some of the intrinsic muscles (adductor pollicis and one head of the flexor pollicis brevis) of the thumb are innervated by the ulnar nerve and long flexor function remains; it is primarily the loss of opposition that poses a problem. A splint that places the thumb in opposition and applied during activity and removed at will, will often suffice to improve function (Figs. 9.12, 9.13, and 9.14).

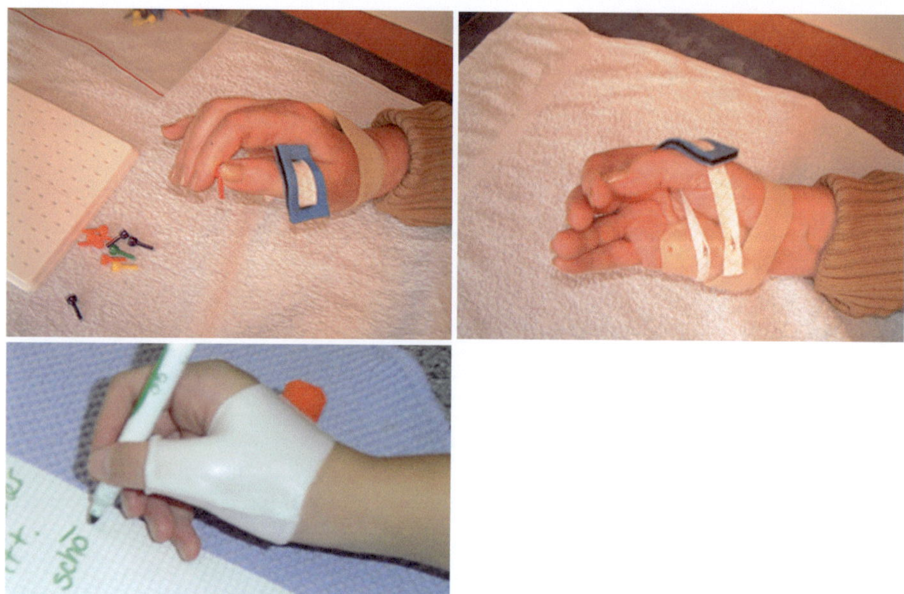

Figs. 9.12, 9.13, and 9.14 Splints to support thumb opposition

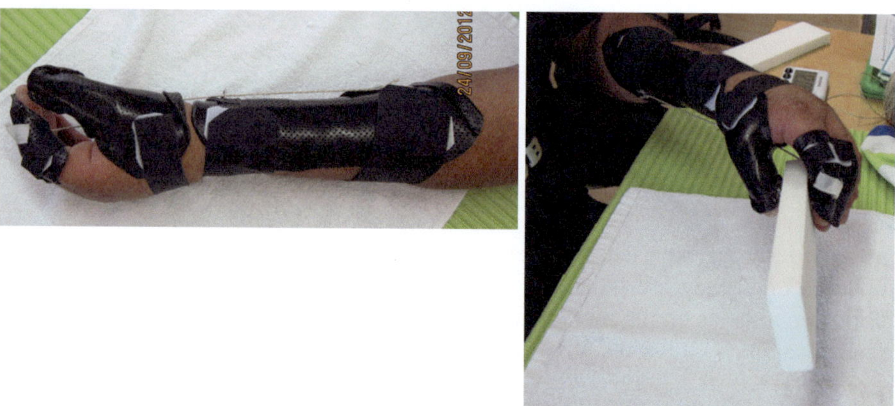

Fig. 9.15 Tenodesis splint for high-level median and ulnar nerve injuries

Splinting for Combined Median and Ulnar Nerve Injuries

When both median and ulnar nerves are injured, a more complex splint is required. The type of splint required depends on the level of the injury; when the level of the injury is in the upper arm and the long flexors are paralyzed, the patient will be able to extend but not flex the fingers; when only extension is possible, custom-made tenodesis splint can provide function (Fig. 9.15)

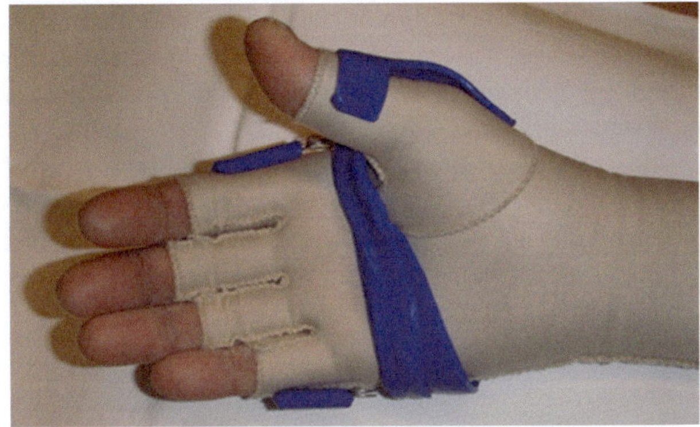

Fig. 9.16 Dynamic splint for combined ulnar and median nerve injury

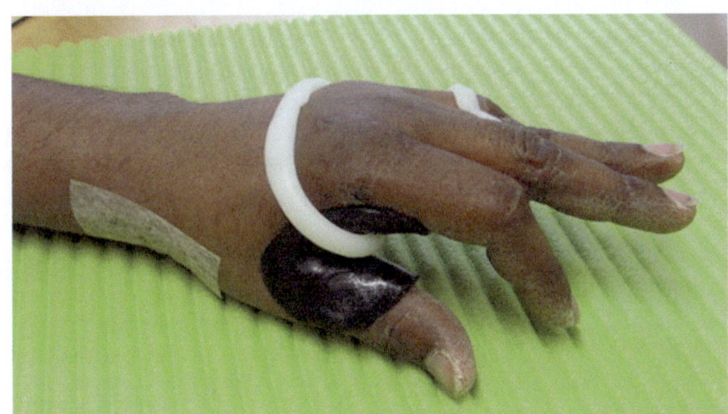

Fig. 9.17 Static splint for combined ulnar and median nerve injury

When the injury occurs in the distal forearm or at the wrist level, a hand-based splint that places the thumb in opposition and prevents hyperextension of the MCP joints can improve function (Figs. 9.16 and 9.17).

Conclusion

Following a peripheral nerve injury, the intensity, quantity, and focus of therapy is highly variable. The nerve that is injured and the resulting impact upon the patient's ability to function play a determining role in the amount and type of therapy that is needed. "Rehabilitation after any nerve repair is slow and may require extensive hand therapy input for up to two years" [26]. The recovery process following a nerve injury can be quite prolonged, and therapeutic treatment must be adjusted regularly; it is important that patients with nerve injuries are reevaluated and treated at regular intervals by a skilled therapist. Treatment may include pain management, treatment of the resultant scar and edema, protective training, interventions to enhance function in activities of daily living, splints to enhance function and minimize contractures, as well as motor and sensory relearning programs. Equally important is the provision of comprehensive home program that educates the patient and allows him or her to participate in their own recovery process. During this process, it is imperative that the patient be supported by the health-care team, and realistic expectations with regard to outcomes are communicated. Although the patient would surely like to hear that he or she can expect a full recovery, in the case of median and ulnar nerve injuries, Vordemvenne et al. [25] found on average that about 70% of hand function could be expected. The health-care team must communicate not only with the patient but with one another with regard to the patient's situation, progress, and further planning.

References

1. Arakawa T, Shigyo H, Kishibe K, Adachi M, Nonaka S, Harabuchi Y. Electrical stimulation prevents apoptosis in denervated skeletal muscle. NeuroRehabilitation. 2010;27:147–54.
2. Berchtold M, Beckmann-Fries V, Meier Zürcher C. Optimal scar management techniques for optimal hand function and aesthetics. IFSSH Ezine. 2016;6(1):19–23. http://www.ifssh.info/ezine.html.
3. Cheng AS, Hung LK, Wong JM, Lau H, Chan J. A prospective study of early tactile stimulation after digital nerve repair. Clin Orthop Relat Res. 2001;384:169–75.
4. Chu M, Chan R, Leung Y, Fung Y. Desensitization of finger tip injury. Tech Hand Upper Extrem Surg. 2001;5(1):63–70.
5. Davis K, Taylor K, Anastakis K. Nerve injury triggers changes in the brain. Neuroscientist. 2011;17(4):407–22.
6. Duff SV, Estilow T. Chapter 45: Therapist's management of peripheral nerve injury. In: Skirven T, Osterman L, Fedorczyk J, et al., editors. Rehabilitation of the hand and upper extremity. 6th ed. St Louis: Elsevier; 2011. p. 620. , 619–33.
7. Elbert T, Rockstroh B. Reorganization of human cerebral cortex: the range of changes following use and injury. Neuroscientist. 2004;10(2):129–41.
8. Elliot D. Surgical management of painful peripheral nerves. Clin Plast Surg. 2014;41(3): 589–613.
9. Ewald S, Wendling W. What does the practice of hand therapy look like in IFSHT member countries? IFSSH Ezine. 2015;5(3):22–5. http://www.ifssh.info/ezine.html.
10. Göransson I, Cederlund R. A study of the effect of desensitization on hyperaesthesia in the hand and upper extremity after injury or surgery. Hand Therapy. 2011;16:12–8.
11. Hundepool CA, Ultee J, THJ N, Houpt P, Research Group 'ZERO', Hovius SER, et al. Prognostic factors for outcome after median, ulnar, and combined medianeulnar nerve injuries: a prospective study. Br J Plast Surg. 2015;68(1):1–8.
12. Lee SK, Wolfe SW. Peripheral nerve injury and repair. J Am Acad Orthop Surg. 2000; 8(4):243–52.
13. Lundborg G, Björkman A, Rosén B. Enhanced sensory relearning after nerve repair by using repeated forearm anaesthesia: aspects on time dynamics of treatment. Acta Neurochir Suppl. 2007;100:121–6.
14. Moscony AMB. Peripheral nerve problems. In: Cooper C, editor. Fundamentals of hand therapy. 2nd ed. Elsevier Inc: St Louis; 2014. p. 272–311.
15. Merzenich MM, Jenkins WM. Reorganization of cortical representation of the hand following alterations of skin inputs induced by nerve injury, skin island transfers and experience. J Hand Ther. 1993;6(2):89–103.
16. Novak CB, von der Heyde RL. Evidence and techniques in rehabilitation following nerve injuries. Hand Clin. 2013;29(3):383–92.
17. Rosén B, Vikström P, Turner S, McGrouther DA, Selles RW, Schreuders TAR, Björkman A. Enhanced early sensory outcome after nerve repair as a result of immediate post-operative relearning: a randomized controlled trial. J Hand Surg. 2015;40E:598–606.

18. Rosén B, Björkman A, Lundborg G. Improved sensory relearning after nerve repair induced by selective temporary anaesthesia- a new concept in hand rehabilitation. J Hand Surg. 2006;31B(2):126–32.
19. Rosén B, Lundborg G. Enhanced sensory recovery after median nerve repair using cortical audio-tactile interaction. A randomised multicentre study. J Hand Surg. 2007;32E(1):31–7.
20. Rosén B, Lundborg G. A model instrument for the documentation of outcome after nerve reapir. J Hand Surg. 2000;25A:535–43.
21. Rosén B, Lundborg G. A new model instrument for outcome after nerve repair. Hand Clin. 2003;19:463–70.
22. Slutsky DJ. Chapter 44: New advances in nerve repair. In: Skirven T, Osterman L, Fedorczyk J, et al., editors. Rehabilitation of the hand and upper extremity. 6th ed. St Louis: Elsevier; 2011. p. 611–8.
23. Spicher C. Handbook for somatosensory rehabilitation. Paris: Saueramps Medical; 2006.
24. Taylor KS, Anastakis DJ, Davis KD. Chronic pain and sensorimotor deficits following peripheral nerve injury. Pain. 2010;151(3):582–91.
25. Vordemvenne T, Langer M, Ochman S, Raschke M, Schult M. Long-term results after primary microsurgical repair of ulnar and median nerve injuries: a comparison of common score systems. Clin Neurol Neurosurg. 2007;109:263–71.
26. Waldham M. Peripheral nerve injuries. Trauma. 2003;5:79–96.
27. Wang Y, Sunitha M, Chung K. How to measure outcomes of peripheral nerve surgery. Hand Clin. 2013;29(3):349–61.
28. Wojtkiewicz DM, Saunders J, Domeshek L, Novak CB, Kaskutas V, Mackinnon SE. Social impact of peripheral nerve injuries. Hand. 2015;10(2):161–7.
29. Yerxa EJ, Barber LM, Diaz O, Black W, Azen SP. Development of a hand sensitivity test for the hypersensitive hand. Am J Occup Ther. 1983;37:176–81.

Peripheral Nerve Tissue Engineering: An Outlook on Experimental Concepts 10

Kirsten Haastert-Talini

Contents

10.1 The Leading Thought

Experimental work on the tissue engineering of peripheral nerves is undertaken with the aim to develop a substitute for the gold standard autologous nerve transplantation to reconstruct longer defects [22] or the alternative approach of suturing a so-called muscle-in-vein graft [49] between the separated ends of a transected nerve. As described in Chap. 5, bioartificial nerve implants, mainly with a single hollow lumen (nerve guidance channels, NGCs), are already available for short defect repair (< 3 cm) but are currently not approved for clinical use in bridging longer nerve defects.

Illustrations: Lena Julie Freund, MSc Animal Biology and Biomedical Science, Aachen, Germany

K. Haastert-Talini, Dr. med. vet.
Adjunct Professor, Institute of Neuroanatomy and Cell Biology, Hannover Medical School, Carl-Neuberg-Str. 1, D-30519 Hannover, Germany

Center for Systems Neuroscience (ZSN) Hannover, Hannover, Germany
e-mail: Haastert-talini.kirsten@mh-hannover.de

© Springer International Publishing AG 2017
K. Haastert-Talini et al. (eds.), *Modern Concepts of Peripheral Nerve Repair*,
DOI 10.1007/978-3-319-52319-4_10

Today, it is obvious that additional modification of hollow NGCs is needed in order to make their properties comparative or even surpassing to those of autologous nerve grafts [14, 22]. Chapter 5 in this book provides an overview of currently available nerve conduits for clinical use. The outer shell of a complex tissue-engineered nerve graft should shield the regenerating nerve tissue from the invasion of scar tissue-forming cells, mainly fibroblasts, and besides of being biocompatible it should allow optimal diffusion of nutrients and metabolites [10, 13]. Biomaterials used for the fabrication of nerve guides are often naturally derived polymers like the extracellular matrix molecules collagen or fibrin, polysaccharides like chitosan, or proteins like silk fibroin [22]. It is highly important that any type of nerve guide provides continuous collapse stability during its degradation *in vivo* to prevent a secondary compression of the regenerated nerve tissue [10]. Tissue engineering approaches further focus on the resembling of the biological structure of autologous nerve grafts as they contain nerve regeneration support cells such as Schwann cells of the repair phenotype [29] that secrete neurotrophic factors and extracellular matrix components. The incorporation of most as possible beneficial properties of an autologous nerve graft into tissue-engineered constructs is warranted [11]. This includes, for most of the experimental approaches, the creation of a three-dimensional endoneurial structure (resembling the bands of Büngner) but also the potential incorporation of other regeneration supporting cells and molecules and provision or induction of an adequate vascularization [14].

10.2 The Characteristics of Peripheral Nerve Regeneration Through Bioartificial Nerve Guides

The properties of a tissue-engineered nerve implant can only be optimized in consideration of the process that naturally occurs when a nerve gap is bridged by a hollow NGC. This process has been evaluated initially using a silicone NGC bridging a 10 mm sciatic nerve gap in the rat [57], and more detailed information has been added over the years [2, 10]. As illustrated in Fig. 10.1, the regeneration process through a hollow NGC with a single lumen includes two main phases. The first period from day 1 to day 14 is characterized by the molecular and cellular phase [2], which is subdivided into the fluid phase, the matrix phase, and the cellular phase [10]. While previous reports mainly focused on the role of migrating perineurial, endothelial, and, most importantly, Schwann cells during this phase [10], there is recently increasing evidence for an important role of the biomaterials potential immunomodulatory properties. The crucial involvement of different phenotypes polarized from invading macrophages during Wallerian degeneration and peripheral nerve regeneration (see also Chap. 1) has been well characterized [8]. The population of the NGC with pro-healing M2 phenotype macrophages has a supportive effect on Schwann cell migration and the following axonal regeneration [41]. Results of our own work indicate, for example, that the nerve guide material chitosan has an immunomodulatory effect and drives the early polarization of invading macrophages especially toward the pro-healing M2c phenotype [52]. The second period, from day 14 onward [2], is characterized

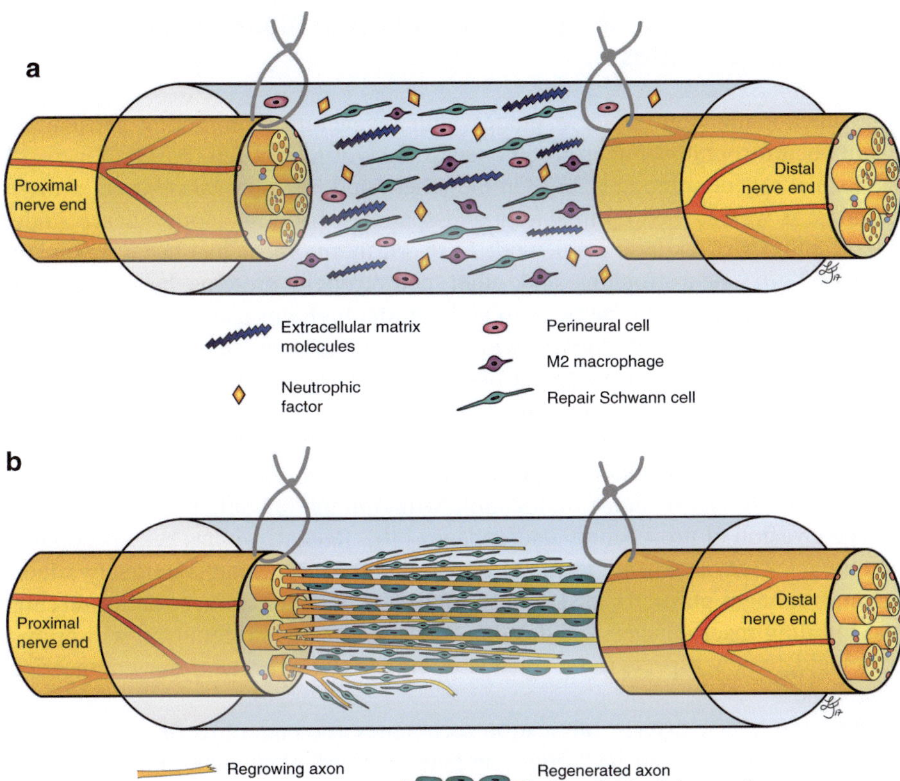

Fig. 10.1 The regeneration process occurring when a peripheral nerve gap has been bridged with a hollow nerve guidance channel can be divided into two main phases (**a**, **b**). (**a**) The molecular and cellular phase: When the nerve guide is initially filled with plasma exudate, extracellular matrix molecules and neurotrophic factors accumulate within its lumen (also referred to as the fluid subphase [10]). This is followed by the formation of a loose fibrin bridge between the separated nerve ends (also referred to as the matrix subphase [10]). The regenerative matrix is then populated by migrating and proliferating Schwann cells, perineurial cells (mainly fibroblasts), and endothelial cells (also referred to as the cellular subphase [10]) and, under optimized conditions, by prohealing macrophages of the M2a and M2c phenotype [41]. (**b**) The axonal and myelination phase: Axonal sprouts travel along the regenerative matrix and the basal laminae provided by the Schwann cells (also referred to as the axonal subphase [10]). Once axonal sprouts have reached their appropriate targets distal to the nerve lesion, they mature and gain in diameter, which induces their myelination by neighboring Schwann cells (also referred to as the myelination subphase [10])

by the ingrowth of axonal sprouts that follow the migrating Schwann cell front into the NGC and later by the myelination of those axons that mature and increase in diameter upon making contact to their appropriate target tissue [10]. Once the nerve defect is healed and the regrown axons have reinnervated their targets, the repair Schwann cells will undergo another reprogramming and adopt the phenotypes of nonmyelinating Remak cells or myelinating Schwann cells [29] again.

10.3   The Concept of Functionalization of Bioartificial Nerve Guides

The functionalization of nerve guides with regeneration-promoting substances, mainly neurotrophins or neurotrophic factors [23, 30], has been proven to increase the axonal regeneration [11, 22]. On the other hand, the optimal mixture of these proteins, as well as their optimal temporal-spatial distribution along a nerve guide, is still subject of experimental work [33]. Furthermore, it has been repeatedly discussed that neurotrophins and other neurotrophic proteins have a limited stability (short half-life time *in vivo*) and that the high dosages, which need to be delivered, can negatively interfere with the regeneration process [10]. Timely and accurate reinnervation of the target tissue could especially be impaired by exaggerating axonal sprouting or entrapment of regenerating axons at sites of high protein concentration [35]. Furthermore, support of the regeneration process may be achieved by supplementing an appropriate cytokine mixture [40] and/or using a biomaterial for nerve guide fabrication with immunomodulatory properties to induce invasion of pro-healing macrophages [36]. Recently, it has been postulated that the microenvironment created after peripheral nerve lesion is uniquely affecting macrophages plasticity [53]. Future studies on the interaction of Schwann cells and invading macrophages, besides those needed for myelin removal, will probably elucidate new aspects to be considered in peripheral nerve tissue engineering approaches.

Over the years, diverse fabrication techniques for preparing structured nerve implants were evaluated including computer-assisted manufacturing, laser-based tissue blotting, and advanced electrospinning or self-assembly of engineered polymers [22]. Furthermore, various surface modifications such as uniform or graduated coating or change of topography ranging from creating an uneven surface up to formation of longitudinal grooves have been evaluated [10, 46].

At first, structural modifications should provide an optimal surface for migrating host cells (so far Schwann cells are mainly considered) and support the remodeling of the bands of Büngner for an optimized axon guidance [11, 27, 39]. This type of modification with physical or topographical guidance cues is also considered to increase the speed and quality of peripheral nerve regeneration [22, 35]. In order to at best resemble the native nerve graft architecture with its axon guiding, elongated, fascicular bands of Büngner, researchers have included fibers (of micro- or nanoscale), filaments, (hydro-)gels, or sponges into the lumen of nerve guides [21, 22, 35].

Secondly, certain modifications should also increase the surface-volume relationship and as such the concentration of regeneration-promoting proteins released from the separated nerve ends and the surrounding into the graft [10]. Other types of modifications can also change the type of nerve implants from being a more or less complex guidance structure or surface for Schwann cell migration into a guiding drug delivery system with pharmacotherapeutic activity [14, 22]. The latter would be of special interest for delayed reconstruction approaches when the initial and physiological upregulation of the repair program and the reprogrammed phenotype of repair Schwann cells [29] have disappeared over time [55]. Pharmacotherapeutic attempts should aim into the incorporation and timely and spatially balanced liberation of neurotrophic molecules, cytokines, or other substances that could modify the phenotype of Schwann cells toward a prolonged support of the regeneration program. It has been demonstrated, for example, that the pharmaceutical substance FTY720P (fingolimod) promotes the phenotype of the repair Schwann cells *in vitro* [26], and it needs future studies to identify if this could be an appropriate pharmacological tool to be incorporated into drug delivering NGCs.

A highly active research field in the last decades has been the potential use of cell therapy in peripheral nerve regeneration. Primary autologous Schwann cells have been representing the first choice of a cellular supplement to tissue-engineered nerve grafts. But although multiple protocols have been developed (e.g., [7, 25, 31, 48]), the harvest of autologous Schwann cells still has almost the same limitation as the harvest of a complete autologous nerve graft. Therefore, the use of autologous Schwann cells as cellular substitute in a clinical setting is still only an option for the most devastating cases [34]. This is the reason why in the recent years, different types of mesenchymal stem cells (derived from bone marrow, Wharton's jelly, adipose tissue, or skin) have been explored as potential substitutes for Schwann cells within tissue-engineered nerve grafts [15, 32, 45, 47]. In this context, the term "tissue engineering" is not limited to the combination of innovative biomaterials with regeneration-promoting cells, extracellular matrix components, or neurotrophic proteins but refers also to the supplementation of these additives to acellular nerve guides to improve also the performance of those in long-distance repair [54]. Figure 10.2 summarizes the main concepts for NGC modification toward the development of an optimized substitute for autologous nerve grafts.

10.4   The Limitations for Translation of Experimental Work into Clinical Application

Although so many innovations arose and so diverse material science approaches have been undertaken, only a very limited number of new products have been translated "from bench to bedside" into a clinical use in the last years. The reasons for this may be as manifold as the engineering ideas.

10.4.1   The Perniciousness of Multidisciplinary Approaches

At first, material scientists may not always be appropriately introduced by their collaboration partners to the specific needs of the biological system they are asked to develop biomaterials for. This can lead to discrepancies, and although the most modern fabrication techniques are used, the developed material compositions may become inappropriate. One example we have experienced in our own group is the

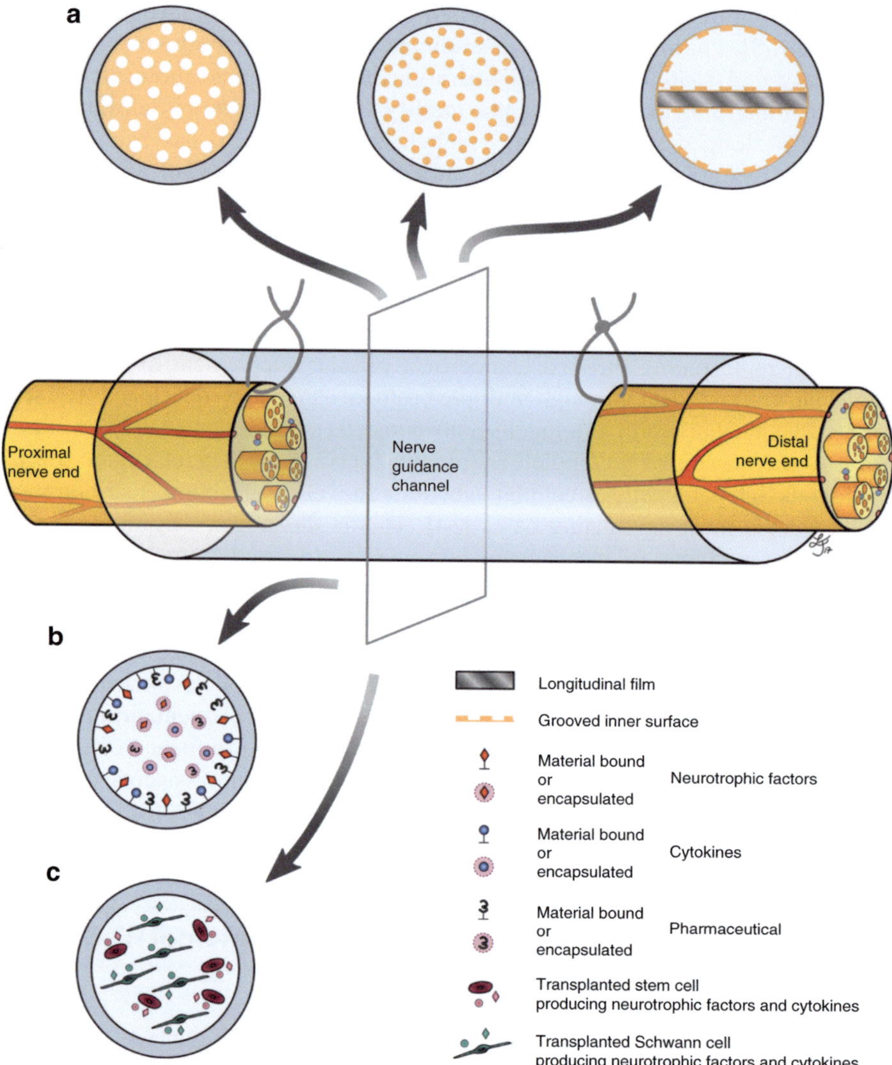

Fig. 10.2 The remodeling of the peripheral nerve regeneration-supporting properties provided by an autologous nerve graft includes three main concepts: (**a**) Providing a three-dimensional guidance structure within the lumen of hollow nerve guidance channels, e.g., by placing a scaffold providing a channel like porosity or another longitudinal guidance structure. This could be achieved, for example, by the use of hydrogels, micro- or nanotubules (*left*), and micro- or nanofibers (*center*) or by the introduction of longitudinal films or grooves (*right*). (**b**) The integration of neurotrophic molecules, cytokines, or pharmaceutical substances that are either bound to the biomaterial or encapsulated for coordinated release should result in the induction of an optimized and prolonged support of the regeneration process. (**c**) The incorporation of support cells aims to provide additional guidance for regrowing axons and to deliver a regeneration-promoting milieu

properties of an electrospun polycaprolactone nerve guide that, although the qualification of this synthetic polymer has been proven before [23], did in our hands induce a massive and deleterious foreign body reaction upon suture between separated nerve ends [12]. Another example is the natural polysaccharide chitosan, which was already in the 1990s used as biomaterial for nerve guide fabrication but demonstrated insufficient stability to be further processed into a clinical product [16]. Anyhow chitosan was further investigated for its use as a biomaterial for neural repair [20], and finally, about 20 years later, an optimized nerve guide was developed [24] and registered for clinical use (nerve gap repair <3 cm) in Europe and the USA (see Chap. 5). A third example, again from our own laboratory, is the use of a hyaluronic acid-based laminin containing hydrogel, originally developed to support regeneration of vascular and neural tissue and with favorable properties for organotypic *in vitro* cultures of sensory dorsal root ganglia [43, 58]. Surprisingly, this specific hydrogel did, instead of supporting the regeneration process, impair peripheral nerve regeneration [38]. In general, hydrogels are used to provide a matrix in which invading or transplanted cells feel as confident as in their physiological destinations. In the context of peripheral nerve regeneration, the hydrogel should support the migration of Schwann cells and other support cells, like, e.g., the pro-regenerative macrophages mentioned before. But when it comes to the questions of how cell-friendly the hydrogel filler and how porous the wall of the NGC should be, specific interactions have to be considered. A recent study demonstrated that when the porous properties of the NGC wall allow uncontrolled infiltration of the cell-friendly hydrogel filler with fibroblasts, this will negatively interfere with the nerve regeneration process [13]. Interestingly, a much simpler modification of a hollow single lumen chitosan nerve guide, namely, the longitudinal introduction of a central chitosan film, was much more successful in increasing the regeneration outcome across a critical defect length of 15 mm in the rat sciatic nerve after immediate [37] and after delayed nerve reconstruction [52]. It can be expected that translation of this two-chambered chitosan nerve guide design into a medical product will be done in a reasonable period. As another example for fruitful collaboration between material scientists and clinical scientists, a more complex, microstructured collagen nerve guide has been developed recently. This specific nerve implant allowed successful nerve repair across a 20 mm sciatic nerve gap in the rat [4] and can also be filled with mesenchymal stem cells for cellular regeneration support [3]. For the latter nerve guide, first results from a clinical study are expected in not later than 1–2 years' time.

10.4.2 The Perils of Preclinical Work

Another obstacle for the translation of all the recent innovations into a promising new medical product, likely, is the fact that many attempts are not adequately evaluated. Simple biocompatibility studies may convince material scientists in the first step, but the novel biomaterials may reveal non-appropriate properties when tested in a more comprehensive way in challenging preclinical models. The highest potential for a propagation of a novel nerve guide into a medical product for clinical use have those approaches that have proven their regeneration-supporting properties in comprehensive *in vitro* and *in vivo* studies. Reports that conclude only from biocompatibility studies with glial cell lines, for example, that a biomaterial is very promising for the fabrication of a NGC, are of minimal value for clinical scientists. *In vitro* studies utilizing cell lines can indeed reveal important information, but the cell lines need to be appropriate for the specific field of neural regeneration addressed. And for a more substantial indication of the biomaterial properties, those studies need regularly to be followed by evaluation of the behavior of primary nerve cells [19]. Also from an ethical point of view, *in vitro* studies are warranted to reduce the number of animals devoted to *in vivo* experiments, but a final comprehensive preclinical *in vivo* evaluation, probably also in different animal models, is unavoidable [19]. A meaningful result from *in vivo* studies, again, will only be obtainable when the different levels of peripheral nerve regeneration that range from proven regeneration of axons across a substantial distance over evidenced reinnervation of distal targets (including indicated specificity of this reinnervation) to, most important, functional recovery [6, 44] are appropriately evaluated. Qualified preclinical animal models should recapitulate the specific nerve regeneration processes, which also take place during peripheral nerve regeneration in human patients. The rat sciatic nerve transection and reconstruction model is the far most used *in vivo* system [1, 18]. Therefore, studies using this model have a higher value with regard to comparability among each other. The rat sciatic nerve model allows evaluation of the reconstruction of defect lengths up to 15–20 mm in a large variety of functional assessments [6, 44]. Although already accepted as a critical defect length model, the critical defect length in human patients is considerably longer (> 10 cm), but the rat sciatic nerve model is still the most appropriate and at best standardized model [1]. The rat median nerve model is considered to be appropriate for extrapolation toward human upper limb or digital nerve injuries and provides from an ethical point of view the high advantage of only minimal impairment for the animal well-being [28]. Rat models have further been developed to additionally address different health conditions and their impact on peripheral nerve regeneration. Studies performed in the Goto-Kakizaki rat that presents with moderately increased and clinically relevant blood glucose levels, thus resembling diabetes type 2 conditions [50, 51], deliver information important for an increasing amount of patients. The mean life expectancy of the laboratory rat additionally allows consideration of delayed nerve reconstruction approaches, a condition that is also highly relevant for a significant number of patients in which primary nerve reconstruction cannot be performed due to a seriously affected general health condition [9]. Finally, after comprehensive preclinical investigation, first inhuman experiences are eventually achievable by repairing the sural nerve after its harvest as an autologous nerve graft with the novel nerve guide that is about to be established for clinical use [5].

For the clinical scientists and clinicians interested into the latest developments and novel medical products that may become available in the nearer future, it is of outmost importance to know about the predictability of the results published from experimental work. Only with this knowledge, the curious reader will be able to conclude that an experimental approach to develop a novel nerve guide is indeed promising and deserves the enterprise to be processed toward a clinical product.

In vitro studies demonstrating the biocompatibility with primary peripheral nerve cells, primary neurons, or organotypic cultures from dorsal root ganglia or spinal cord preparations and also providing evidence for a stimulated neurite outgrowth from primary sensory or motor neurons by the tested biomaterial provide a legitimization for *in vivo* studies [19]. For this legitimation, it is not obligatory to demonstrate optimal biomaterial properties by complex peripheral nerve regeneration *in vitro* models, which evaluate neurite outgrowth from spinal cord slices to peripheral nerve segments or nerve guide material [17, 56]. The latter *in vitro* systems are the most refined and sophisticated ones, but on the opposite, they are difficult to establish and to propagate.

Preclinical *in vivo* models should then be chosen in consideration of ethical as well as assessment concerns [18, 44]. When used as a stand-alone readout, histomorphometrical evaluation of axonal regeneration across a certain distance does not predict the final functional outcome [6]. Such evaluation needs to be combined with at least one predictive assessment of functional recovery, which can be the histological proof of target tissue reinnervation by retrograde labeling or the recording of evoked compound muscle action potentials upon stimulation of the repaired nerve [6]. Diverse readouts exist to determine the recovery of complex voluntary motor functions (e.g., video gait analysis in sciatic nerve models or grasping test in median nerve model) as well as sensory recovery (e.g., von Frey mechanical pain threshold test) [6, 44]. Such complex evaluations are of course of highest value for the translation of study results into the propagation of a novel approach toward a medical product for clinical use.

There is one examination technique, which should receive specific consideration in this context: the calculation of the Sciatic Functional Index (SFI) from measured distances between the spread toes of the hind paws. The index indicates complete sciatic nerve dysfunction (no spreading of toes possible anymore) when calculation of its mathematical formula equates to -100, and it indicates full sciatic nerve function when the formula equates to zero [44]. Alternative calculation of sciatic nerve trunk specific indices, like for the tibial nerve or peroneal nerve (TFI, PFI), is also performed related to the fact that regenerating axons could be misdirected from the common sciatic nerve stem into the non-appropriate trunk [6]. Although results for the calculation of the different functional indices are still often presented, the assessment of the same has considerable drawbacks when used to determine the degree of functional motor recovery after nerve gap reconstruction procedures. While demonstrating enough reliability and values returning to approximately pre-injury values after nerve crush injuries of different severity [42], obtainable values after repair of nerve transection, especially with any type of gap bridging implant, are often not valid to discriminate between treatments as they do not reach values indicating substantial recovery [44]. This is due to the already mentioned innervation of alternative motor targets by misdirected regenerating axons leading to contractures, joint stiffness, and abnormal paw posture in general [6, 44]. Therefore, conclusions about the quality of novel NGCs, which are drawn on the basis of SFI or alternative calculations and histomorphometrical analysis of regenerated axons close to the implant, should not be considered as strong and predictive for a clinical translatability.

10.4.3 The Regulatory Constraints

Finally, before a novel medical product is made available for clinical use, there are regulatory aspects that have to be considered and which represent a considerably high burden especially for cell-supplemented tissue-engineered products that are not supplemented with the patient's autologous cells, the so-called advanced therapy medicinal products (regulation on advanced therapies (Regulation (EC) 1394/2007)). Furthermore, regulatory work including toxicity and biocompatibility tests as well as phase I and II clinical studies are expensive to perform already for novel nerve implants that are probably cell-free but composed of completely novel biomaterials. Consideration of the last two points underscores the preceding paragraphs from this chapter and the need to design experimental studies with high predictability of a potential clinical value of novel tissue-engineered nerve implants.

References

1. Angius D, et al. A systematic review of animal models used to study nerve regeneration in tissue-engineered scaffolds. Biomaterials. 2012;33:8034–9.
2. Belkas JS, et al. Peripheral nerve regeneration through guidance tubes. Neurol Res. 2004;26: 151–60.
3. Boecker AH, et al. Pre-differentiation of mesenchymal stromal cells in combination with a microstructured nerve guide supports peripheral nerve regeneration in the rat sciatic nerve model. Eur J Neurosci. 2016;43:404–16.
4. Bozkurt A, et al. Efficient bridging of 20 mm rat sciatic nerve lesions with a longitudinally micro-structured collagen scaffold. Biomaterials. 2016;75:112–22.
5. Bozkurt A, et al. The proximal medial sural nerve biopsy model: a standardised and reproducible baseline clinical model for the translational evaluation of bioengineered nerve guides. Biomed Res Int. 2014;2014:121452.
6. Brushart TM. Outcomes of experimental nerve repair and grafting. In: Brushart TM, editor. Nerve repair. New York: Oxford University Press; 2011. p. 159–95.
7. Casella GT, et al. Improved method for harvesting human Schwann cells from mature peripheral nerve and expansion in vitro. Glia. 1996;17:327–38.
8. Chen P, et al. Role of macrophages in Wallerian degeneration and axonal regeneration after peripheral nerve injury. Acta Neuropathol. 2015;130:605–18.
9. Dahlin LB. The role of timing in nerve reconstruction. Int Rev Neurobiol. 2013;109:151–64.
10. Daly W, et al. A biomaterials approach to peripheral nerve regeneration: bridging the peripheral nerve gap and enhancing functional recovery. J R Soc Interface. 2012;9:202–21.
11. Deumens R, et al. Repairing injured peripheral nerves: bridging the gap. Prog Neurobiol. 2010;92:245–76.
12. Duda S, et al. Outer electrospun polycaprolactone shell induces massive foreign body reaction and impairs axonal regeneration through 3D multichannel chitosan nerve guides. Biomed Res Int. 2014;2014:835269.
13. Ezra M, et al. Porous and nonporous nerve conduits: the effects of a hydrogel luminal filler with and without a neurite-promoting moiety. Tissue Eng Part A. 2016;22:818–26.

14. Faroni A, et al. Peripheral nerve regeneration: experimental strategies and future perspectives. Adv Drug Deliv Rev. 2015;82-83:160–7.
15. Faroni A, et al. Adipose-derived stem cells and nerve regeneration: promises and pitfalls. Int Rev Neurobiol. 2013;108:121–36.
16. Freier T, et al. Biodegradable polymers in neural tissue engineering. In: Mallapragada S, Narasimhan B, editors. Handbook of biodegradable polymeric materials and their applications. Stevenson Ranch: American Scientific Publishers; 2005. p. 1–49.
17. Gerardo-Nava J, et al. Spinal cord organotypic slice cultures for the study of regenerating motor axon interactions with 3D scaffolds. Biomaterials. 2014;35:4288–96.
18. Geuna S. The sciatic nerve injury model in pre-clinical research. J Neurosci Methods. 2015;243:39–46.
19. Geuna S, et al. In vitro models for peripheral nerve regeneration. Eur J Neurosci. 2016;43:287–96.
20. Gnavi S, et al. The use of chitosan-based scaffolds to enhance regeneration in the nervous system. Int Rev Neurobiol. 2013;109C:1–62.
21. Gonzalez-Perez F, et al. Extracellular matrix components in peripheral nerve regeneration. Int Rev Neurobiol. 2013;108:257–75.
22. Gu X, et al. Neural tissue engineering options for peripheral nerve regeneration. Biomaterials. 2014;35:6143–56.
23. Gu X, et al. Construction of tissue engineered nerve grafts and their application in peripheral nerve regeneration. Prog Neurobiol. 2011;93:204–30.
24. Haastert-Talini K, et al. Chitosan tubes of varying degrees of acetylation for bridging peripheral nerve defects. Biomaterials. 2013;34:9886–904.
25. Haastert K, et al. Human and rat adult Schwann cell cultures: fast and efficient enrichment and highly effective non-viral transfection protocol. Nat Protoc. 2007;2:99–104.
26. Heinen A, et al. Fingolimod induces the transition to a nerve regeneration promoting Schwann cell phenotype. Exp Neurol. 2015;271:25–35.
27. Hodde D, et al. Characterisation of cell-substrate interactions between Schwann cells and three-dimensional fibrin hydrogels containing orientated nanofibre topographical cues. Eur J Neurosci. 2016;43:376–87.
28. Jager SB, et al. The mouse median nerve experimental model in regenerative research. Biomed Res Int. 2014;2014:701682.
29. Jessen KR, Mirsky R. The repair Schwann cell and its function in regenerating nerves. J Physiol. 2016;594:3521–31.
30. Jiang X, et al. Current applications and future perspectives of artificial nerve conduits. Exp Neurol. 2010;223:86–101.
31. Keilhoff G, et al. Neuroma: a donor-age independent source of human Schwann cells for tissue engineered nerve grafts. Neuroreport. 2000;11:3805–9.
32. Khuong HT, et al. Skin derived precursor Schwann cells improve behavioral recovery for acute and delayed nerve repair. Exp Neurol. 2014;254:168–79.
33. Klimaschewski L, et al. The pros and cons of growth factors and cytokines in peripheral axon regeneration. Int Rev Neurobiol. 2013;108:137–71.
34. Levi AD, et al. The use of autologous schwann cells to supplement sciatic nerve repair with a large gap: first in human experience. Cell Transplant. 2016;25:1395–403.
35. Marquardt LM, Sakiyama-Elbert SE. Engineering peripheral nerve repair. Curr Opin Biotechnol. 2013;24:887–92.
36. McWhorter FY, et al. Physical and mechanical regulation of macrophage phenotype and function. Cell Mol Life Sci. 2015;72:1303–16.
37. Meyer C, et al. Chitosan-film enhanced chitosan nerve guides for long-distance regeneration of peripheral nerves. Biomaterials. 2016a;76:33–51.
38. Meyer C, et al. Peripheral nerve regeneration through hydrogel-enriched chitosan conduits containing engineered Schwann cells for drug delivery. Cell Transplant. 2016b;25:159–82.
39. Mobasseri A, et al. Polymer scaffolds with preferential parallel grooves enhance nerve regeneration. Tissue Eng Part A. 2015;21:1152–62.
40. Mokarram N, Bellamkonda RV. A perspective on immunomodulation and tissue repair. Ann Biomed Eng. 2014;42:338–51.

41. Mokarram N, et al. Effect of modulating macrophage phenotype on peripheral nerve repair. Biomaterials. 2012;33:8793–801.
42. Monte-Raso VV, et al. Is the Sciatic Function Index always reliable and reproducible? J Neurosci Methods. 2008;170:255–61.
43. Morano M, et al. Nanotechnology versus stem cell engineering: in vitro comparison of neurite inductive potentials. Int J Nanomedicine. 2014;9:5289–306.
44. Navarro X. Functional evaluation of peripheral nerve regeneration and target reinnervation in animal models: a critical overview. Eur J Neurosci. 2016;43:271–86.
45. Oliveira JT, et al. Bone marrow mesenchymal stem cell transplantation for improving nerve regeneration. Int Rev Neurobiol. 2013;108:59–77.
46. Rajaram A, et al. Strategic design and recent fabrication techniques for bioengineered tissue scaffolds to improve peripheral nerve regeneration. Tissue Eng Part B Rev. 2012;18:454–67.
47. Ribeiro J, et al. Perspectives of employing mesenchymal stem cells from the Wharton's jelly of the umbilical cord for peripheral nerve repair. Int Rev Neurobiol. 2013;108:79–120.
48. Rutkowski JL, et al. Purification and expansion of human Schwann cells in vitro. Nat Med. 1995;1:80–3.
49. Schiefer JL, et al. Comparison of short- with long-term regeneration results after digital nerve reconstruction with muscle-in-vein conduits. Neural Regen Res. 2015;10:1674–7.
50. Stenberg L, Dahlin LB. Gender differences in nerve regeneration after sciatic nerve injury and repair in healthy and in type 2 diabetic Goto-Kakizaki rats. BMC Neurosci. 2014;15:107.
51. Stenberg L, et al. Nerve regeneration in chitosan conduits and in autologous nerve grafts in healthy and in type 2 diabetic Goto-Kakizaki rats. Eur J Neurosci. 2016;43:463–73.
52. Stenberg L, Stössel M, et al. submitted. Regeneration of long-distance peripheral nerve defects after delayed reconstruction in healthy and diabetic rats is supported by immunomodulatory chitosan nerve guides.
53. Stratton JA, Shah PT. Macrophage polarization in nerve injury: do Schwann cells play a role? Neural Regen Res. 2016;11:53–7.
54. Szynkaruk M, et al. Experimental and clinical evidence for use of decellularized nerve allografts in peripheral nerve gap reconstruction. Tissue Eng Part B Rev. 2013;19:83–96.
55. Tajdaran K, et al. An engineered biocompatible drug delivery system enhances nerve regeneration after delayed repair. J Biomed Mater Res A. 2016;104:367–76.
56. Vyas A, et al. An in vitro model of adult mammalian nerve repair. Exp Neurol. 2010;223:112–8.
57. Williams LR, et al. Spatial-temporal progress of peripheral nerve regeneration within a silicone chamber: parameters for a bioassay. J Comp Neurol. 1983;218:460–70.
58. Ziv-Polat O, et al. The role of neurotrophic factors conjugated to iron oxide nanoparticles in peripheral nerve regeneration: in vitro studies. Biomed Res Int. 2014;2014:267808.

Index

A

Allodynia, 104, 110, 115, 116

Axon

    axonal injury, 9

    axonal outgrowth, 8

    axonal regeneration/regenerating axon, 5,
7–9, 39, 45–47, 68, 92, 128, 130, 135

    axonal sprouts, 8, 9, 129

    myelinated, 1, 2, 7, 91

    unmyelinated/non-myelinated, 1

B

Bandaging, 114

Bands of Bungner/Bands of Büngner, 8, 128, 130

Biomaterial, 81, 85, 93, 128, 130–136

Brachial plexus injury

    neonatal brachial plexus palsy (NBPP)

        cookie test, 67

        postoperative evaluation, 73–75

        preoperative evaluation, 66–70

        towel test, 67

    (nerve) root avulsion

        postganglionic, 20, 21, 66

        preganglionic, 20, 66

        rupture, 66, 68, 69

C

Clinical diagnosis, 11, 22

Clinical follow-up

    International Classification of Functioning,
Disability and Health, 75, 111

    multifactorial evaluation, 111

Complications

    bleeding, 103–104

    infection, 104

    nerve damage during surgery, 102

    postoperative, 103

    scar, 104

    swelling, 85, 113, 114

    vascular injury, 103

    wound dehiscence, 104

Connective tissue

    endoneurium, 2, 3, 8

        interfascicular, 2

    epineurium

        epineurotomy, 6

        external, 42

        internal/interfascicular, 2, 6

    gliding, 2

    mesoneurium, 42

    perineurium, 2

Contracture, 118–120, 123, 135

Cortical representation, 113, 115, 116

Cytokine

    cytokine mixture/liberation of cytokines,
130, 131

    cytokine pro-inflammatory, 8

D

Decompression, 56

    decompression of the carpal tunnel, 55–57

Desensitization, 116

Disability of arm, shoulder and hand (DASH),
102, 111, 115

Dressing, 115

E

Edema, 45, 113, 114, 123

Electrodiagnostic techniques

    compound muscle action potential (CMAP)

        low, 15, 16

        normal, 15

© Springer International Publishing AG 2017
K. Haastert-Talini et al. (eds.), *Modern Concepts of Peripheral Nerve Repair*,
DOI 10.1007/978-3-319-52319-4